A-LEVEL
AND AS-LEVEL

GEOGRAPHY

David Burtenshaw

Longman

LONGMAN A AND AS–LEVEL REFERENCE GUIDES

Series editors: Geoff Black and Stuart Wall

TITLES AVAILABLE
Biology
Chemistry
English
Geography
Mathematics
Physics

Longman Group UK Limited,
Longman House, Burnt Mill, Harlow,
Essex CM20 2JE, England
and Associated Companies throughout the world.

© Longman Group UK Limited 1990

First published 1990

British Library Cataloguing in Publication Data

Burtenshaw, David
  Geography.
  1. Great Britain. Secondary schools. Curriculum
  subjects: Geography
  I. Title
  910.71241

  ISBN 0–582–06388–4

Designed and produced by
The Pen and Ink Book Company,
Huntingdon, Cambridgeshire.
Set in 10/12pt Century Old Style.

Printed in Singapore

ACKNOWLEDGEMENTS

I wish to extend my sincere thanks to all my colleagues who have tolerated my obsession with the geographical alphabet and not least Paul Farres and John McClatchey who assisted with some of the entries. Murray Thomas, my advisor, continued to offer constructive help and advice which has made him such a valued examiner. Geoff Black and Stuart Wall, the series editors, have listened patiently and offered constructive criticism throughout the project. My family, especially Helen, heard the moans and groans sympathetically while the home computer listened poker-faced to my curses when my typing skills failed yet again under the combined onslaught of Christmas festivities, Biogeography entries and student essays competing for time. The omissions, slips and errors cannot be blamed on such crises and events; they are mine alone or the product of the necessity to squeeze the factual overload of A-level and AS-level into the page constraints of this small volume. Perhaps you, the revising student, may dream of log cabins or tropical coconut palms when the task gets boring – I do! So I dedicate these notes to the revising dreamers, whatever your dream and wherever you are.

HOW TO USE THIS BOOK

Throughout your A-level and AS-level course you will be coming across terms, ideas and definitions that are unfamiliar to you. The Longman Reference Guides provide a quick, easy-to-use source of information, fact and opinion. Each main term is listed alphabetically and, where appropriate, cross-referenced to related terms.

- Where a term or phrase appears in **different type** you can look up a separate entry under that heading elsewhere in the book.
- Where a term or phrase appears in **different type** and is set between two arrowhead symbols ◄   ►, it is particularly recommended that you turn to the entry for that heading.

ABLATION ZONE

That area of a **glacier** or **ice sheet** where more ice and snow is being lost through evaporation and **meltwater** than is being gained from the incoming flow of ice. It is normally the lower areas of a glacier although the area where ablation is taking place varies with the season. It is divided from the **accumulation zone** by the **equilibrium line**.
◀ Glacier ▶

ABRASION

The effect of the mechanical process by which materials in **transport** grind, polish and scratch the surface over which they are passing. The materials can be boulders in a stream or **glacier**, pebbles in a stream or wind-borne sand. **Corrasion** is the process of erosion of the surface. Abrasion is responsible in part for the hollowing of the glacial **cirque**.

ACCESSIBILITY

The ease with which a place may be reached from other places which can be measured in **network theory**. This term is generally used in relation to access by industry to raw materials and of workers to work. It is also used to describe the ability of people to reach the places where goods and services are administered in the urban system.

ACCUMULATION

The process by which toxic substances concentrate in the higher levels of a food chain (see Fig. A.1). It is worse in aquatic food chains where there are more **trophic levels**. The best case of accumulation is the concentration of DDT in the higher levels of the food chain. DDT was used as a pesticide but it has gradually been washed into streams where the food chain has gradually accumulated it so that the upper carnivores have a dangerous concentration of the pesticide. **Minimata disease** is another case of food chain accumulation.

The word may also be used to describe the progressive acquisition of wealth to reinvest in further capital accumulation within the capitalist economic system. It is a term used in Marxist analyses of the development of capitalism.

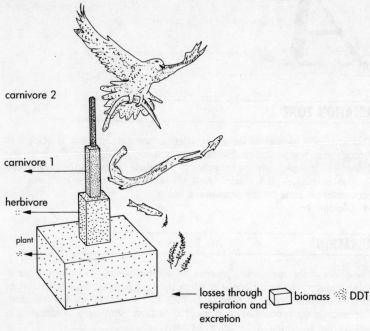

Fig. A.1 The accumulation of DDT in a foodchain

ACCUMULATION ZONE

The area of a **glacier** or an **ice sheet** where there is a net gain in snow and ice, and where the snow is transformed into ice. It also refers to the part of a slope where there is an increase in the amount of material so that the level rises. It is associated with the **foot slope**.
◀ Glacier ▶

ACID RAIN

Acid rain or acid deposition refers to the people-made increase in acid brought about by air pollution. It is found in southern Scandinavia, North-West Europe, Eastern Canada, and North-East USA, although some of these areas are not the sources, for example, Norway and Sweden. It is trans-frontier pollution.

Burning of fossil fuels releases sulphur dioxide (SO_2) and nitrogen oxides

(NO_x) which react with sunlight and ozone, and dissolve to produce acid rain. (In UK 70 per cent of acid rain is caused by sulphur products and 30 per cent by nitrogen.) Power stations (61 per cent of SO_2 in UK), domestic central heating and car exhausts are the main sources.

ACTIVE POPULATION

The people of working age in a population, normally 16–64. **Working population** is a preferred term because it refers to those in work and those temporarily unemployed and is a better measure of the **labour force**.

ACTIVITY RATE

The proportion of a population that is employed in paid work. It can be subdivided by sex or even age groups. The figure is not favoured by many because it excludes work in the home which supports the working partner.

ADDITIONALITY

◄ European Regional Development Fund ►

ADVANCED SOCIETY

◄ Mobility transition model ►

AFFORESTATION

◄ Forest ►

AGE–SEX PYRAMID

A graphical representation of the numbers or proportion of each age cohort of males and females in a population. A pyramid can be drawn for any area ranging from a country to a city or even to a smaller area such as a census **enumeration district**. The overall shape of the pyramid can be related to the **demographic transition model**. The pyramids in Fig. A.2 show a **constructive** or youthful **pyramid** for England and Wales in 1881, Stage 2, and a **regressive pyramid**, Stage 4, in 1981. Pyramids also indicate how a population changes through time. A single pyramid such as that for Cologne (Fig. A.3) can reveal much about the effect of major events such as wars and recessions on the evolving structure of a population.

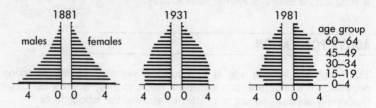

Fig. A.2 The demographic transition illustrated by the population of England and Wales

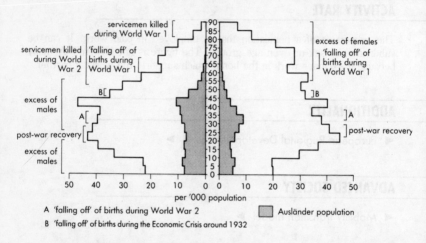

A 'falling off' of births during World War 2

B 'falling off' of births during the Economic Crisis around 1932

Ausländer population

Fig. A.3 The population of Cologne, by age and sex, 1980

AGE SPECIFIC BIRTH RATE

◀ Birth rate ▶

AGE SPECIFIC DEATH RATE

◀ Death rate ▶

AGE SPECIFIC FERTILITY RATE

◀ General fertility rate ▶

AGGLOMERATION ECONOMIES

The economies obtained by a factory or other economic production unit which are derived from its location in close proximity to other like or linked economic activities. The similar production units will be in a spatial cluster or agglomeration.

AGRICULTURAL GEOGRAPHY

The description and explanation of the ways in which the patterns of agricultural activity vary from place to place. The main focus is upon land use and the description and explanation in physical (climate and soils) and human (technology, machinery and labour) terms. Other factors do come into consideration such as land tenure, labour supply, farm size and farm fragmentation besides economic variables which affect the inputs and outputs. Today, some agricultural geography looks at the role of agriculture in the economic development of the **developing world**. The role of political intervention in agriculture is increasingly important and therefore the study of agricultural intervention may range from the **Common Agricultural Policy** of the **European Community** to the effect of a **green belt** on agriculture.

AID

This is flow of resources from **developed** to **developing countries** and includes the flow of resources in the form of capital, technology and expertise, credits for exports from the developed country to the developing country and education and training assistance. Aid may be **bilateral** in that is is from one country to another, e.g. Great Britain to Zimbabwe or it may be **multilateral**, i.e. from a group of countries or an organisation representing a group of countries such as the **European Community** or the World Bank. Charitable organisations also distribute aid, e.g. Oxfam. Aid is often politically motivated and can be seen as **neocolonial** in that it is sent to former colonies rather than other more deserving cases. Aid is frequently **tied** so that the recipient must use the aid to purchase goods or services from the donor country. Some cases of tied aid will not benefit the recipient because the donor has insisted on them buying goods which are inappropriate to their needs. **Food aid** can alter tastes and be to the ultimate disadvantage of the local food producers. **Untied aid** is the ideal but it needs careful monitoring to ensure that it is spent wisely. Some believe that too much aid can result in **aid dependency**, a growing reliance on especially food aid for survival, which may not be the fault of the recipients.

AID DEPENDENCY

◀ Aid, Intermediate technology ▶

AIR MASS

Air which remains relatively stationary over an area of uniform character begins to assume the temperature and humidity characteristics of that surface whether it be ocean or the earth's crust. A large area of air may acquire these relatively homogeneous values which are particularly found in areas dominated by large **anticyclones**. These areas of uniform characteristics are known as air masses and they are known by their area of origin latitudinally; **Arctic, polar, tropical** and either **maritime**, or **continental** according to the underlying environment.

AIRPORT

The area designed for the use of passenger and freight aircraft transport and its interchange with other modes of transport. There is a hierarchy of airports in a country from the small private airfield up to the major international airport. Airports are major users of capital and land and major employers both directly and indirectly. The interchange facilities with other modes are also space extensive.

ALIENATION

The general feeling of hopelessness which affects those whose power in the social system is little or absent. It is regarded as affecting the poorer members of society most although many would suggest that the process may affect middle class professionals, e.g. teachers, if society undervalues their contribution to well being.

ALLUVIAL FAN

A fan-shaped area of deposition deposited by a stream when its velocity decreases as it emerges from a valley onto a plain or level area. Fans exist where small streams leave former **hanging valleys** in glaciated areas to join the main valley. They are more common in desert areas and spread intermittently as the result of the intermittent nature of the **flash floods**. The finer materials are deposited furthest from the apex of the fan. When several fans merge in an arid area they are called a **bahada** (bajada).

ALONSO'S BID RENT THEORY

This model was developed to explain variation in land values, land use and intensity of land use in cities. It built upon the ideas developed by **Von Thünen** and the basic criteria of **accessibility** and **transport** costs. Alonso assumed that all workplaces were at the city centre (see Fig. A.4.) and therefore households would have to pay to reach their workplace. The further they lived from work, the greater the cost of accessibility and the less money available for land and housing. He assumed that all households wanted as much land as possible so the rich could live in the more plentiful, cheaper land on the city fringe whereas the poor would live on less land close to work. Each land use has its own bid rent and when these are superimposed, it is possible to see which uses might dominate land use in each zone from the city centre. When the curves for individual uses are transferred onto a surface, then it does give credence to the **Burgess model**. Other uses have been substituted on the curves so that they can show, for example, the house type zones extending out from the inner city. Today, most cities have a much more complex land value surface more akin to the spider's web in Fig. A.5 which is really a refinement of the model. The affluent can also reside near the city centre because they can trade off land cost against distance costs especially if they do not value gardens. This can lead to **gentrification**. Also the curves today tend to peak towards the outside of the city at key locations e.g. close to a motorway junction which might be valued by both industry and retailing.

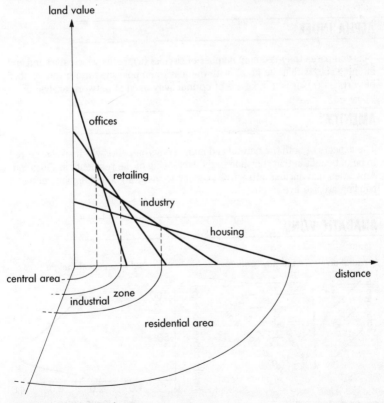

Fig. A.4 Alonso's bid-rent diagram

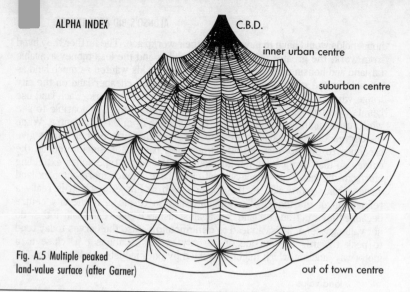

C.B.D.

inner urban centre

suburban centre

Fig. A.5 Multiple peaked
land-value surface (after Garner)

out of town centre

ALPHA INDEX

This compares the maximum number of circuits (u), paths which start and end
in one place, within the graph with the maximum possible for a given number
of vertices (v). It is a measure of **connectivity** used in network analysis.

AMENITY

The aspects of both the natural and human environments which are perceived
to be of benefit in that they add to the well-being of the residents in a location.
Amenities may include attractive scenery to visit, a good shopping centre, a
housing area to live in etc.

ANABATIC WIND

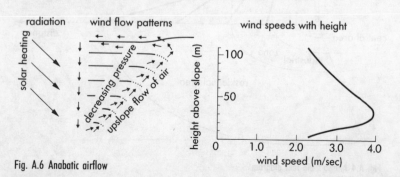

radiation

wind flow patterns

wind speeds with height

solar heating

decreasing pressure

upslope flow of air

height above slope (m)

wind speed (m/sec)

Fig. A.6 Anabatic airflow

An upslope wind caused by the greater insolation received on valley sides which heat up causing air to rise; this then draws more air up the valley, as shown in Fig. A.6.

ANAEROBIC

Conditions in the soil without free oxygen present; associated therefore with waterlogging and hence gleying.

ANASTOMISING CHANNEL

◄ Braiding ►

ANTICYCLONE

A term first used in the 1860s to denote an area of high atmospheric pressure with a circular form to its isobars and whose characteristics contrast with those of the cyclone or frontal depression.

APARTHEID

The policy of racial segregation of different ethnic groups which was first developed in South Africa in 1948. Each ethnic group must live in a specific area although the boundaries between the areas for some of the ethnic groups have become blurred in recent years. The most significant development of the policy has been the founding of Bantustan homelands, e.g, Transkei, for the indigenous black population. The whole policy ensures the continued social and economic domination of South African society be the white minority. Jobs, marriage and the location of a person's home are all controlled by strict regulations. The first major steps towards the dismantling of the system were announced in February 1990 by President F. W. de Klerk.

APPLIED GEOGRAPHY

The application of geographical knowledge and techniques to the solution of economic and social issues and problems. Such studies are mainly carried out by geographers working in conjunction with public authorities such as planners, the water boards and, increasingly, private sector companies willing to sponsor research which might give them some financial advantage. Any study which has practical benefits for society may be called applied and this tends to include most research in geography today.

ARCH

Normally formed where waves have cut through a headland (see Fig. A.7) exploiting lithological weaknesses by **differential erosion** and leaving the arch which eventually collapses to form a **stack**. Durdle Roor in Dorset is an arch.

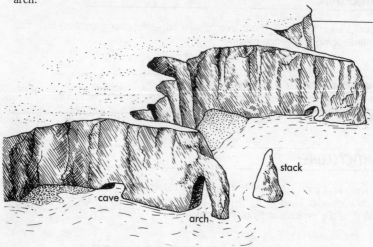

stack

cave

arch

Fig. A.7

ARCTIC AIR MASS

Air whose source is in the artic which gives rise to very cold winter weather and cold snaps in spring and summer. Because it passes over oceans on its way to the UK it is unstable and often contains showers of snow, sleet, and hail.

ARCTIC–DESERT TYPE

◀ Population-resource ratio ▶

AREAS OF OUTSTANDING NATURAL BEAUTY

The National Parks and Access to the Countryside Act 1949 allowed the Countryside Commission to designate AONBs. About 11% of England and Wales has been designated including the Isles of Scilly (16km^2) and the North Wessex Downs (1738km^2). Planning authorities have to consult the Countryside Commission on all aspects of development in an AONB.

ARITHMETIC MEAN

◀ Mean ▶

ARROYOS

The dried up bed of a gully in the Americas and Spanish speaking world. (It could refer to the dried up floor of a **wadi**). It can contain an **ephemeral stream**.

ARTESIAN WELL

A well in which the water rises naturally to the surface under pressure from water contained in the underlying permeable stratum when that stratum rises above the level of the well head elsewhere in the structural basin. The London Basin forms the ideal conditions for artesian wells because the chalk (which rises up to the North Downs and Chiltern Hills) has a higher water table thus providing the necessary pressure for water to be forced upwards in the wells of central London. Wells lowered the level of the water in the basin until recent years when the closing of many of the old wells has caused the water table to rise again beneath the city.

ARTIFICIAL LEVEE

The heightening of a **levee** to prevent flooding from a river. Such works often result in increased danger to nearby areas from the low frequency, high magnitude floods that overflow the levee. These cause greater damage than the higher frequency, low magnitude floods which may have preceded the building of the levee.

ASSIMILATION

◀ Core frame concept ▶

ASSISTED AREAS

A general term used to refer to all the areas which received assistance from regional policy, i.e. the **special development areas, development areas** and **intermediate areas**.

ASSOCIATION

In biogeography this term refers to a grouping of plants in a particular **ecological niche** such as an **oasis**. The association will have several dominant yet interdependent species.

ASYMMETRICAL VALLEY

A valley whose cross-profile shows that the slopes on one side are steeper than those on the other side. The processes which give rise to asymmetry are not fully understood but they include **periglacial** processes in an environment where the sunnier south facing slope was subject to greater **denudation**.

ATMOSPHERE

One of the three major constituents of the earth's environmental system through which water moves as a part of the **hydrological cycle** returning to the **lithosphere** and **hydrosphere**.

ATMOSPHERIC CIRCULATION

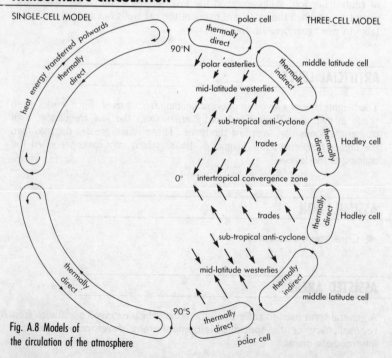

SINGLE-CELL MODEL

THREE-CELL MODEL

heat energy transferred polwards

thermally direct

90°N

polar cell

thermally direct

middle latitude cell

polar easterlies

mid-latitude westerlies

thermally indirect

sub-tropical anti-cyclone

trades

thermally direct

Hadley cell

0°   intertropical convergence zone

trades

thermally direct

Hadley cell

sub-tropical anti-cyclone

mid-latitude westerlies

thermally indirect

middle latitude cell

90°S

thermally direct

thermally direct

polar cell

Fig. A.8 Models of the circulation of the atmosphere

The basic model of circulation is shown in Fig. A.8 and contains three cells. Between the cells at high altitude there exists jet streams: high speed westerly winds (easterly in the tropics) which affect the movement of the cells and cause waves in the upper westerlies called Rossby waves. These in turn affect the medium-term location of the major pressure systems or cells.

ATOMIC POWER

The production of commercially usable nuclear energy; converting the heat generated by splitting the atom into electric energy. It provides 3.9% of the energy used in the world and a significantly higher proportion in those countries in the **developed world** which are short of **fossil fuels**, e.g. France. Further expansion of nuclear power is unlikely because of the high costs and the safety fears surrounding generation and disposal of spent fuel.

ATTRITION

The reduction in size of materials in **transport** as a result of friction between the particles. It takes place during transport by wind, wave and water.

AUTOTROPHS

Plants which have the capability to use carbon dioxide as the main source of carbon and to gain their energy from the sun's radiation. Autotrophs are able to manufacture their own sources of food and are not dependent on outside food sources.

AVALANCHE

The fast descent of a mass of ice, rock and snow from a mountainside where it is either partially thawed and unstable, or recent and uncompacted. They are a threat to the life of recreationalists such as skiers and may threaten settlements, especially if the forest protection is removed around a mountain village. In some areas **acid rain**, which has killed the trees, has exposed villages to an increased threat from avalanches. In regions where the avalanche threat is greatest tracks have been mapped and procedures enforced to reduce the risk e.g. the firing of avalanche guns to dislodge precarious ice with the help of the shockwaves.

It is a form of **environmental hazard**.

BACKWASH

A term used by **Myrdal** and in the **growth pole theory** for the movement of people, jobs, investment and resources to the core region. It is the negative effect of development in the core and the opposite of the **spread effect**.
◀ Growth pole theory, Myrdal's model of cumulative causation ▶

BAHADA

◀ Alluvial fan ▶

BANK CAVING

◀ Load, Meander ▶

BARCHAN

A crescent-shaped dune which has two horns which extend down wind. The crescent slope on the windward side is gentler than the slip face on the leeward side. Wind **deflation** and **saltation** carry the sand up the windward slope and push the particles over the crest to fall down the steeper leeward slope. Barchans are normally found in groups all of which are aligned to the wind.

BAR CHARTS

Diagrams (see Fig. B.1) which are used when there is one quantitative scale showing the data. The height of the bar represents the frequency. The horizontal scale is based on divisions by place, names or other categories suitable to the data. These can show composite data.

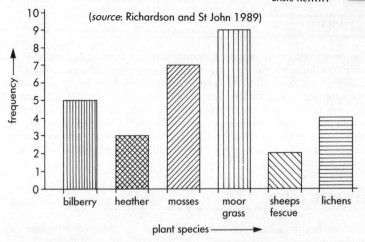

Fig. B.1 A bar chart showing the results of a quadrat survey carried out on a stretch of open moorland in Yorkshire

BARLOW COMMISSION REPORT 1940

The founding white paper for post-war regional policy in Great Britain was called 'The Distribution of the Industrial Population'. It proposed measures to assist the establishment of industry in depressed areas and to decongest the south-east. It also proposed **new towns**. It laid the foundations for policies which were to use a 'carrot and stick' method to alter the distribution of economic activity with the **Distribution of Industry Act 1945**.

BASAL SLIDING

The movement of a **glacier** over its floor and the weight of the glacial ice which results in **internal deformation**, i.e. the ice behaves like a fluid due to its weight. The speed of slide is also influenced by the nature of the bedrock.

BASE FLOW

◀ Hydrograph ▶

BASIC ACTIVITY

Economic activities which are founded upon demand from outside of the region of production. It is also called **export base activity**.

BATHOLITH

The largest **intrusive volcanic** landform which is composed of granite and, unlike the **laccolith**, has no observable base to the intrusion. Very often several of the batholiths are linked at depth to form a larger sub-surface structure known as a **pluton**. The granite batholiths of Devon and Cornwall are part of such a larger structure. Most batholiths are surrounded by metamorphic rocks where the heat of the **magma** has altered the rocks, e.g. limestone to marble. Batholiths often contain areas of non-ferrous minerals such as tin and copper and, when exposed, give rise to landscapes which have tourist potential.

BAY HEAD BEACH

A small **beach** which occupies a bay that is recessed between two headlands.

BEACH

The store of sediments which exists between the high and low water marks (see Fig. B.2). It extends beyond the high water mark due to the effects of

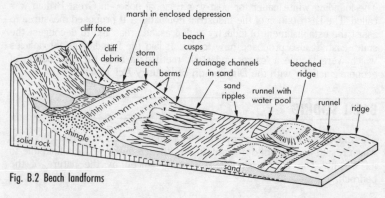

Fig. B.2 Beach landforms

deposition extending the beach. The beach can be seen as a system and is composed of several landforms.
◄ Beach cusps, Berms, Runnels, Sand ripples, Storm beach ►

BEACH CUSPS

Crescent shaped deposits of sand and shingle which are found on a beach. They are formed by swash and backwash; their frequency and size are related to the time interval between the waves and the nature of the waves, e.g. spilling waves will lead to larger cusps.

BEDLOAD

The coarser material carried along the bed of a stream by the force of the streamflow. The energy of the stream will determine the size of the particles being moved. They are mainly rolled, pushed or bounced by **saltation**, therefore the largest particles are moved only when the velocity is highest, i.e. in **flood**.

BEHAVIOURAL GEOGRAPHY

This is an approach to the study of human geography which looks at the way in which people comprehend, react to and modify their environment on the basis of their own cognitive processes. It arose as a reaction to **spatial analysis** and the **scientific method** with its notion of *rational economic people*. It recognised that people do not have perfect knowledge and that all decisions are made on the basis of imperfect knowledge. People are **satisficers** in that they make decisions which suit them best. Behavioural studies have been used in the study of agricultural innovation, industrial location, housing and where to shop.

BENELUX

The economic union established in 1948 between Belgium, the Netherlands and Luxembourg; it can be considered to be a forerunner of the **European Community**.

BENETTONISATION

A term used to describe the method of organisation of the production and selling of clothing developed by Benetton. Manufacturing is carried out in small establishments in 'the Third Italy', i.e. central and north Italy but the colours are not dyed into the fabrics until computer data on the most popular colours and styles is received. The shops retailing the finished products are **franchised** so that all that the company actually does is to design the products and supervise the computer operations between sales and manufacturing. It is an approach to manufacturing and services which relies on small firms.

BENIOFF ZONE

◀ Subduction zone ▶

BERMS

A ridge on a beach which represents the accumulation of shingle at the last high tide level. There will often be several berms because of the variable height of the tides. They are also called **beach ridges**

BETA INDEX

A simple measure of **connectivity** which measures the ease with which movement is possible between one vertex to another: $\beta = e/v$

BID RENT THEORY

◀ Alonso's bid rent theory ▶

BIFURCATION RATIO

A statistic used to define the form of the drainage network which states the relationship between the streams at one order and those at the next highest level.

BIG BANG

The popular term given to the deregulation of the financial markets which brought new firms to the City of London in the period after 1987. Some see it as the cause for increases in the demand for office space and further **office decentralisation** while others see the influx of new highly qualified labour migrants as the cause of house price inflation in the late 1980s.

BILATERAL AID

◄ Aid ►

BIOGAS

Methane derived from animal dung: it ferments in pits from which the gas is collected. There are 7 million such **intermediate technology** plants producing biogas in China.

BIOGEOGRAPHY

The study of the distribution of plants and animals on the earth's surface. The interrelationships between these and the broader environment is the subject of **ecology**.

BIOMASS

The total volume of living organisms which can be supported at a particular **trophic level**. Biomass is measured in terms of its calorific value or in terms of its dry weight.

BIOMASS ENERGY

Energy resources such as timber, dung and straw. Twenty-one countries depend on biomass energy for three-quarters of their energy supplies. In the Sahelian countries of Burkina Faso and Mali, 90% of the energy is supplied by timber. Biomass energy is the source of fuel for cooking in much of the **Developing World**.

BIOME

A major climax community of which there are ten in the world according to most authorities (as shown in Fig. B.3). Within each biome there is a fairly uniform set of plants and animals in terms of their adaptation to their environment. Most biomes approximate to the major climatic regions of the world (see Fig. B.4). Mountain areas are excluded from the classification because, depending on their location, mountain areas contain a varied set of biomes associated with different altitudinal zones.

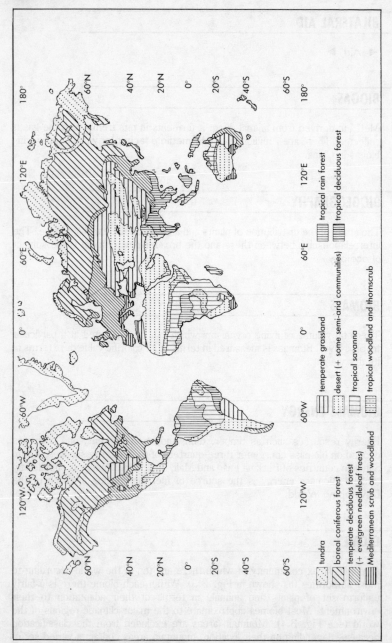

Fig. B.3 Major terrestrial biomes of the world

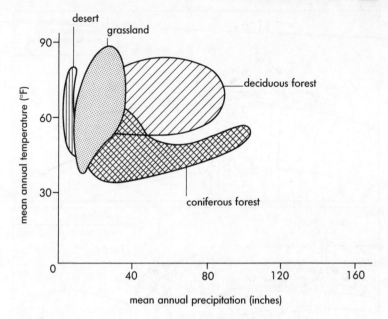

Fig. B.4 Distribution of the six major biomes in terms of mean annual temperature and mean annual rainfall in inches (courtesy of National Science Foundation)

BIOSPHERE

The zone in which all life on earth exists. It is adjacent to, and overlaps, the lithosphere and atmosphere. It is here that matter is transferred from the hydrosphere to the atmosphere.

BIOTECHNOLOGIES

Scientific developments which have enabled agriculturalists to gain greater output from the same unit area. Developments include: hydroponics; new plant varieties which give higher yields such as IR8 or Miracle Rice; technologies that control light so flowers bloom for Christmas.

BIRD'S ANYPORT MODEL 1971

A model, illustrated in Fig. B.5, which attempts to show the broad pattern of port development over time. There are three stages: a) the primitive port adjacent to firm ground where early docks were excavated; it was sited close to a bridging point of the river from which quays extended down river and

BIRD'S ANYPORT MODEL

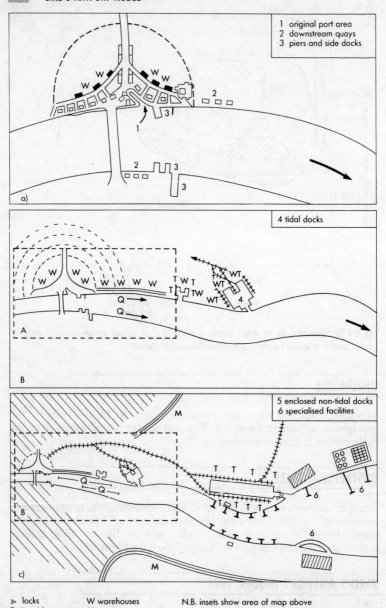

a)
1 original port area
2 downstream quays
3 piers and side docks

b)
4 tidal docks

c)
5 enclosed non-tidal docks
6 specialised facilities

➤ locks W warehouses N.B. insets show area of map above
Q riverside quays M motorways
T transit sheds

Fig. B.5 Bird's Anyport model

across the river becoming more elaborate all the time; b) the creation of wet tidal docks downstream in deeper water with more onshore facilities such as warehouses and transit sheds; c) linear quays in enclosed, non-tidal docks to accommodate larger vessels followed by the building of special quays for example, roro ferries, **containerisation** and oil terminals. Some now suggest a fourth stage: the closing of upstream facilities and the redevelopment of redundant quays for marinas, housing and new centres of employment.

BIRDSFOOT DELTA

◀ Distributaries ▶

BIRTH RATE

◀ Crude birth rate ▶

BLACK EARTH

◀ Chernozem ▶

BLACK ECONOMY

Also known as the informal sector, this term refers to those **service** sector activities which are not regulated by law and which are subject to cash payment to avoid taxation. It is sometimes associated with the underclass and jobs such as nannying, domestic cleaning and unregulated child minders. Some black economy activities take place in legitimate firms merely by the non-recording of payments, in order to make profits appear smaller and thus reduce tax liability.

BLIZZARD

Snowfalls in storm conditions when the wind whips the snow into drifts which then cut off settlements and immobilise transport. When a blizzard hits Chicago, the whole of the air transport system of the USA is affected because O'Hare airport is the hub of the American air transport system.

BLOCKING ANTICYCLONE

An anticyclone which drifts away from its area of origin and establishes itself in high latitudes diverting the tracks of frontal **depressions** to the north and south. In summer, the drift north of a blocking high from the Azores will result in prolonged drought periods, e.g. 1989. In winter, the westward extension of

BLOWHOLE

A vertical hole in a cliff extending from the surface at the top of the cliff down to a cave at sea level. Water forced into the cave by the waves is forced up the shaft of the blow hole slowly widening the hole. It is also known as a geo.

BORDER

The common term for the boundary between two countries which is officially termed the frontier.

BOREAL CONIFEROUS FOREST BIOME

The forest is found in a broad belt across North America and Eurasia where there are long, cold winters and short summers. The forest biome contains evergreen conifer trees with needle leaves whose shape enables them to shed snow. They are normally found in large stands of a dominant species. The stands have very sparse lower layers and the ground is covered with lichens, moss and bog moss depending on the light conditions. The species are few in number and their productivity is 35% of that of the tropical rain forest because of the low energy inputs. The litter is slow to decay and much of the nutrient store is here or in the biomass. The areas are also characterised by podsol soils. Adaptation takes place as a result of burning caused by lightening strikes which light the flammable litter layer. People start fires by careless behaviour. The forest is slow to recover because of the low winter temperatures which prevent plant growth. Clearance from mining, settlement, transport and agriculture has reduced the area of forest although it is the commercial exploitation of timber for paper pulp and construction that once threatened the biome. Today forests in the boreal belt are subject to management to ensure that the resource is not depleted. Acid rain seems to be a further threat to the forest, killing new shoots and needles and slowing growth rates. Other hazards to the biome exist in some areas, e.g. fall out from the Chernobyl accident has produced mutations to the forest plants in those areas close to the former reactor where fallout was heaviest.

BOSERUP'S THEORY

Writing in 1965, Boserup suggested that a society only developed when it was threatened by lack of resources. In this way people respond to population pressure by discovering new agricultural methods and crop strains. To her, necessity is the mother of invention and therefore resource provision rises to meet population growth.

BOURNE

◀ Ephemeral stream ▶

BRAIDING

The process by which a stream breaks up into a series of channels. It is due to the decrease in discharge resulting in the selective deposition of the load as shingle and gravel bars. These divide the stream into channels which join and redivide. It also occurs in deltas.

BRAIN DRAIN

◀ Labour migration ▶

BRANCH PLAN ECONOMY

The descriptive term given to the types of industrial and service activities which have been developed in the north and west of Great Britain and in the peripheral and development regions of many countries. The plants are branches or subsidiaries of manufacturers whose main production and control is either elsewhere in the country, normally the core region, or overseas. The South Wales economy has become a branch plant economy due to the growth of plants such as Ford at Bridgend and those of several Japanese manufacturers such as Sony.

BRANDT COMMISSION

The popular term used to name the commission headed by the former West German Chancellor Willy Brandt which began work in 1977 as the Independent Commission on International Development Issues. Its members were drawn from all over the world (except communist countries) and they produced two reports; *North-South: A Programme for Survival 1980* and *Common Crisis North-South; Co-operation for World Recovery 1983*. The reports stress the interdependence of the world's countries and suggest how a transfer of resources should take place from North to South, how the South could increase food production and how poverty could be reduced. The second report was occasioned by the failure of countries to take action and so it recommended financial, trade, energy and food policies to assist in the development of the South.

BRAUN-BLANQUET RATING SYSTEM

A method of vegetation mapping which describes both the degree of **cover** or occurrence and the grouping or distribution of species.

Cover
+ = sparse, cover small
1 = plentiful, but cover small
2 = numerous, cover greater or = 5%
3 = any number, cover 25–50%
4 = any number, cover 50–75%
5 = covering greater than 75% of area

Grouping
Soc.1 = isolated individuals
Soc.2 = grouped or tufted
Soc.3 = patches or cushions
Soc.4 = colonies or carpets
Soc.5 = pure population

BRAZIL TYPE

◄ Population–resource ratio ►

BREAKERS

Waves approaching the shore oversteepen and break. Breakers are subdivided into *spilling* which break over a distance, *plunging* which break in a crashing form down onto the beach and *surging* which neither plunge nor spill but surge up the beach. The movement of the water up the beach is the *swash* whereas the movement back down the beach is the **backwash**. If the swash is stronger then the waves are said to be **constructive**: these are produced by spilling and surging breakers. If the backwash is stronger then the waves are said to be **destructive**: these are produced mainly be plunging waves. Both swash and backwash interact in a complex fashion to produce *rip currents*, very strong flows of water running at right angles out from the shore which take the excess water from the waves back out to sea. Waves are refracted or bent into the shape of the coastline by reaching the shape of the submarine shallowing at headlands before that in bays.

BRICKEARTH

◄ Wind deflation ►

BROWN SOIL (EARTH)

Soils with a mull or moder horizon but having no B horizons of accumulation or depletion of clay or sesquioxides. The brown A horizon merges into a lighter coloured horizon of alteration (B) (Cambic horizon). Thought to be the equilibrium soil of the temperate forests. Calcareous and acid sub-orders of this type also exist.

BUCKET SHOP

A term which has come into common use to describe travel shops which retail cut-price air tickets. They flourish in the large cities and in those areas of the city where the foreign, casual tourists tend to stay, e.g. Earls Court, London.

BURGESS MODEL

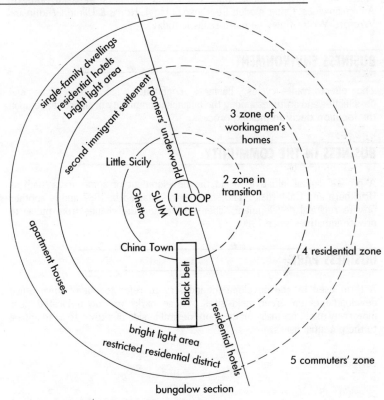

Fig. B.6 The Burgess model applied to 1920 Chicago

This model, illustrated in Fig. B.6, is also known as the **concentric zone model** and was first proposed in 1925 following studies of the city of Chicago by the Chicago school of urban sociology. Using ecological theory applied to people such as the concepts of **invasion** and **succession**, and social survey methods, they tried to interpret the social and economic forces which were shaping the slums to see how they influenced the social organisation of the people who lived there. To obtain the pattern of rings of development they mapped juvenile delinquency, for example, and surveyed the activities of gangs. Therefore they saw the city structure as a series of rings of growth all being fed by the growth of the central area which attracted more people to work in the city. As each new group came to the city they clustered in areas with their fellow immigrants, e.g. Little Sicily and then, as they made their way in life, they moved outwards invading the next zone. The invading social groups were then in a dominant position in the social order of the next zone. Like plants, people were seen to be competing for the best locations. **Hoyt** said that the zones did not take sufficient account of communications. The model was a product of its time when large numbers were migrating to American cities. Other models i.e. those of **Hoyt, Harris & Ullman, Mann** and **Wreford-Watson** may be more realistic today.

BUSINESS ENVIRONMENT

The effect trade cycles, business confidence, sales projections and possibilities and attitudes among the business community in an area have upon the **location decision making** process.

BUSINESS IN THE COMMUNITY

A private capital initiative to aid the development of some areas such as Blackburn and Calderdale. The scheme was started in 1987 and is another private regional development scheme reflecting the change from public to private initiatives since 1980.

BUSINESS PARK

A term used by the development industry to refer to out of town office developments on green field sites. These parks are often located near motorway exits, normally on the more desirable side of a town so as to appeal to the potential users.

CALCICOLE

A plant which requires a lime rich soil.

CALCIFICATION

The process, or processes, of soil formation in which the soil is supplied with an excess of free calcium which saturates all sites in the exchange complex.

CALCIFUGE

A plant which cannot tolerate lime in the soil and flourishes in acid soils.

CALDERA

◄ Volcano ►

CAPILLARY RISE

The process by which water can rise up a soil profile towards the surface. This results from a reduction in water content at the surface producing a potential gradient. The capillary potential created is in excess of the gravitational potential and the water therefore moves along the direction of the total potential gradient. In this way water at depth is 'attracted' towards the surface against the force of gravity.

CAPITAL RESOURCES

The buildings, machinery, equipment and tools that a society possesses in order to support itself. These may also be called **cultural resources**.

CARBONATION

A form of **chemical weathering** found mainly in **karst** areas in which rainwater and CO_2 form a weak carbonic acid (H_2CO_3) which dissolves away limestones as calcium bicarbonate.

CARBONIFEROUS LIMESTONE

The geological name for the lowest and oldest subdivision of the Carboniferous rocks in Britain. It is not solely a limestone but the limestone with its massive blocks gives rise to the distinctive **karst** scenery found in the Mendips, parts of the Pennines and the Brecon Beacons besides the Burran in Western Ireland and the classic karst area of west Yugoslavia.

CARNIVORES

The organisms at the third and subsequent **trophic levels**, as shown in Fig. C.1, which obtain their energy and food by consuming the **herbivores**. They

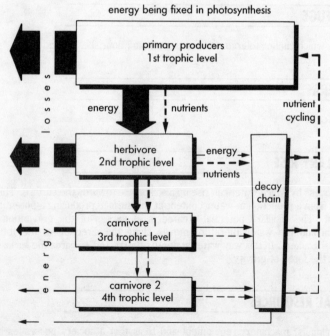

Fig. C.1 An ecosystem: energy flow and material cycling

are flesh eaters and, at the top trophic level, there is the human race. Carnivores are also known as *secondary consumers*

CARTEL

An organisation to control the production and marketing of a product so that the maximum profit is obtained due to agreements between the producers to restrict competition. **OPEC** (the Organisation of Petroleum Exporting Countries) is the most famous cartel whose activities affect trade patterns. Each producer has a specific output which has been agreed by the cartel and so prices and profits are increases due to supply matching demand. The 1973 oil price rise introduced by OPEC gave rise to the mid-1970s world recession. Cartels lose their effectiveness if one member steps out of line or if other producers are not members. OPEC is no longer so effective because of political differences among its members and because exporting countries such as Great Britain are not members.

CATCHMENT BASIN SYSTEM

A model of the flow of energy in the form of water, matter in the form of sediments and matter in solution passing through a river basin and its subsystems, the **slope system** and the **channel system** as shown in Fig. C.2.

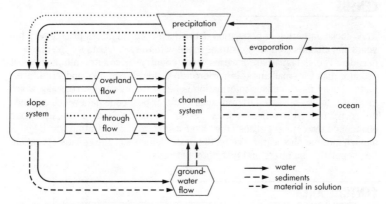

Fig. C.2 The catchment basin system

The outputs of this system are the water and material transferred back to the oceans. As a result of the operation of the processes of **overland flow**, **through flow** and **groundwater flow** the throughput alters the nature of the catchment basin. However, the return of water to the oceans enables solar energy to evaporate the water and so provide the energy input for the system to maintain itself. The catchment basin or **drainage basin** is the fundamental system through which the **denudation** process operates.

CATENA

A repeated sequence of soils of the same age, derived from similar parent materials whose differences result from variations in relief and drainage.

CATION EXCHANGE CAPACITY (CEC)

The sum total of exchangeable cations that the soil can absorb per unit volume of soil resulting from the net negative charges found on the the clay-humus complex. The CEC is expressed in milliequivalents per 100 grams of soil.

CAVE

A cavity in the face of a cliff which has been exploited by the forces of marine erosion on areas which are lithologically weaker and subject to **differential erosion**. It may develop into a *geo* as a result of a combination of **abrasion** and **hydrological** action. The term also refers to a chamber beneath the surface of the earth often found in areas of **karst** scenery and formed by the **solution** of the limestone although the development of the cave normally owes more to other agents of erosion. See Fig. S.5.

CENSUS

The official gathering of population data which normally takes place every ten years and enables comparisons to be made with the same data for the previous decade. The data gathered varies from country to country although most produce data for small units, (enumeration districts), parishes and larger areal units. Today much census information is being recorded electronically which enables it to be converted to grid square data or post code data which can then be correlated with other sources. Some countries, e.g. Denmark, have abandoned the census because they have a system of registration which keeps a much more sophisticated eye on the basic elements of the census operation, i.e. population numbers and migration patterns.

CENTRAL AREA

A broader term than **central business district** used to convey the feeling of all the uses in the heart of a city, commerce, retailing, recreation, government and residential.

CENTRAL BUSINESS DISTRICT (CBD)

This is an American term used to describe the focus of the commercial and civic activities of a city. The CBD has grown up at the heart of the historic city, at the focus of the transport routes where accessibility is greatest. It is dominated by office, retail and entertainment uses besides being the focus of local government buildings. In European cities there may also be a large residential population besides other anachronistic buildings which are part of the city's heritage, e.g. cathedral and castle. The larger the CBD the more the functions within it segregate both horizontally and vertically. It has grown and its actual location may shift over time (**zone of assimilation**) and leave behind areas (**zone of discard**). The CBD's spread may be constrained by relief, a river, parkland, governmental buildings, a ring road, railway lines or planning zoning. It has a definite structure of land uses which are open to mapping and measurement of association such as **nearest neighbour**, and *centre of gravity*. Activities can be distinguished by floor level. It has been delimited by the *central business height index* – total CBD floor space divided by the total ground floor space for a street block. A value of 1 or more indicates the CBD. The *central business intensity index* is the product of the central business floor space divided by the total floor space times 100. A figure of 50% or more indicates a CBD block. Rateable values have been used and divided by the front meterage of the use as an indicator of the value of each site in the CBD. The CBD is surrounded by the **zone in transition** or the frame of the **core frame** model.

CENTRALISED STATE

A country whose government takes control over most of the affairs of its citizens and has no intermediate tier between the national government and local government. It normally results in the centralisation of both political and economic control as is the case in both London and Paris.

CENTRAL RECREATION DISTRICT

The area in the **central area** of a town or city which is dominated by recreational activity. It might be a zone of theatres and cinemas, or a zone of museums; alternatively it could be an area of restaurants and night clubs and, in some cities, a red-light district.

CENTRAL TENDENCY

A way of describing a whole set of data which is synonymous with the average or **arithmetic mean**.

CHANNEL

The gutter-like form which contains a stream or a river. In arid or semi-arid areas the channel may be dry. The shape of this channel is known as its **channel variables**. It normally contains water which has reached it from precipitation, **groundwater flow, throughflow** and **overland flow** and is flowing towards the sea or an inland sea or lake. In some Fenland channels the flow may be reversed. Flow is also reversed in the tidal estuary.

CHANNEL GEOMETRY

◄ Channel variables ►

CHANNEL GRADIENT

◄ Channel variables ►

CHANNEL STORAGE

A stage in the **hydrological cycle** referring to the water stored in stream channels on its course to the oceans or to the atmosphere following evaporation.

CHANNEL SYSTEM

A sub-system of the **catchment basin system** in which the characteristics of the stream channel alter.

CHANNEL VARIABLES

width (w) — measured at water surface

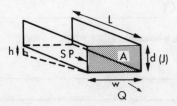

Fig. C.3 The channel variables

depth (d)	– varies, normally expressed as mean depth across section (J)
cross-section area (A)	– Jw or mean depth x width
wetted perimeter (P)	– length of channel margin in contact with flowing water P = 2d + w
channel gradient (S)	– change in elevation per unit of length
velocity of flow (V)	– distance travelled per unit of time
discharge (Q)	– volume of water passing through a cross-section per unit of time.

CHEMICAL WEATHERING

A group of chemical reactions which lead to the breakdown or decay of rocks.
◀ Carbonation, Hydrolysis, Oxidation, Reduction, Solution ▶

CHERNOZEM

Old term used in zonal classification of soils, still used in the FAO system. Soil with a thick black, or near black, organic matter rich A horizon (Mollic A); high in exchangeable calcium; calcium carbonate accumulation in lower horizons. Characteristic soil in a cool subhumid climate under a vegetation of tall and midgrass prarie.

CHI^2 TEST χ^2

A statistical test which tests whether the populations are identical in a significant form. It is a **non-parametric** test because the data is **nominal**, i.e. placed in a series of categories and **ordinal**, i.e. the data is ranked. It compares the observed frequencies (O) with a set of expected frequencies (E) to see if the difference is a matter of chance or of sampling error.

CHOROPLETH MAP

A map such as that in Fig. C.4 where the shading patterns indicate the size of the data values for each aerial unit. The divisions between the shading values has been predetermined by other statistical methods such as **mean** and standard **deviations** or by regular intervals.

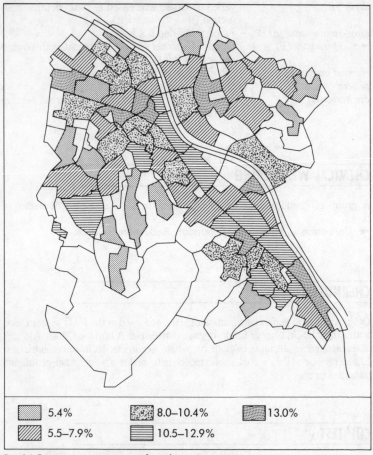

Fig. C.4 Foreigners as a percentage of population, Bonn 1983

Legend:
- 5.4%
- 5.5–7.9%
- 8.0–10.4%
- 10.5–12.9%
- 13.0%

CHRISTALLER'S CENTRAL PLACE THEORY

This theory, first developed in Germany in 1933, attempts to explain the size and pattern of settlement in an urban system, see Fig. C.5. The model assumes that:

- there is an isotropic surface
- transport costs vary with distance
- places all vary in the type and quality of goods and services which they supply
- the most efficient arrangement of central places on an isotropic surface is a lattice of equilateral triangles which can then form a mesh of hexagons

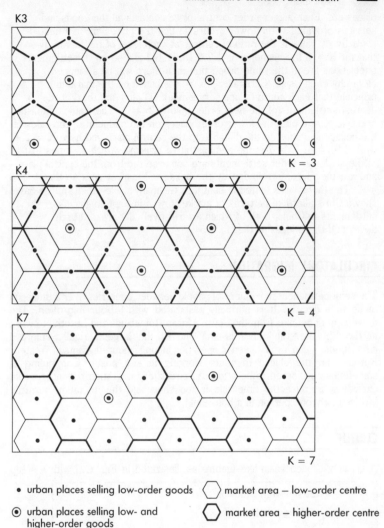

- urban places selling low-order goods ⬡ market area – low-order centre
- ⊙ urban places selling low- and ⬡ market area – higher-order centre
  higher-order goods

Fig. C.5 Central place theory

- people will use the nearest place supplying a good or service
- each place will provide a range of services.

Different retailing and service functions differ in the range and **threshold** of the goods which they supply and these places can be grouped into places with similar characteristics. Those centres supplying the lowest order goods have the lowest range and threshold and form the densest network, those supplying higher order goods form the next network with fewer places etc. Each higher order central place contains all the goods and services provided at the lower orders so that the places are in a nested hierachy. K value is the ratio of centres at each level of the hierarchy so that for K = 3 the sequence is 1,2,6,18,54,162 and so on. If settlements' populations are in the same proportion then there will be a stepped hierarchy of city size. Christaller's K = 3 is based on the marketing principle that each central place serves itself (1) plus one third of the six nearest settlements of the next order below (2) = 3. The traffic optimising principle K = 4 puts the lower order places on the midpoint of the sides of the hexagon to shorten the distance to the central place and here K = 1 + (6 × 0.5) = 4. K = 7 is the administrative principle which assumes that all of the six lower order settlements are administered from the central place and are therefore surrounded by the hexagonal market area so that 1 + 6 = 7. The theory was tested in southern Germany while other studies have shown Christaller-like patterns of market towns in East Anglia and settlements in China, both of which are relatively flat areas. Lösch developed these ideas further in 1954.

CIRCULATORY MIGRATION

The semi-permanent movement of people between countries over a timespan of up to a decade. It is normally associated with labour migration. The movement of peoples from the South African Homelands and Lesotho to work in the Witwatersrand area is still one of the largest, rigidly enforced circulations; it can also be seen in operation among labour migrants in Western Europe and the skilled migrants working in key posts. Commuting is sometimes considered to be a form of circulatory migration but most would exclude it because the timescale is too short and the migration normally involves only one person in a household.

CIRQUE

A glacially eroded basin-like feature as illustrated in Fig. C.6 with a steep

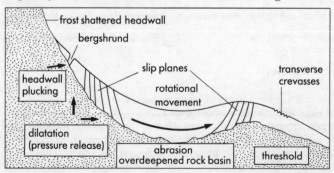

Fig. C.6 The cirque

headwall, formed by **plucking**, which is found at the head of a glacial valley. Some may still contain cirque **glaciers** although most are armchair hollows which may contain a small lake or **tarn**.

◄ Abrasion ►

CITY ACTION TEAMS

Founded in Nottingham and Leeds in 1988 these teams are a further private initiative to foster economic development.

CITY GRANTS

Begun in 1988, these are a government scheme to replace urban development grants and urban regeneration grants. They form a major part of the **urban programme**.

CITY RICH, CITY POOR SECTIONS

These are the terms given by **Lösch** to the products of his attempt to rotate a series of K networks of central places so that the outcome was more akin to the real world. The pattern of higher order places which results does show sectors of higher order and sectors of lower order places radiating from the first order settlement. This is closer to the reality of the world than Christaller's central place theory.

CLAY-HUMUS COMPLEX

The soil is made up of inorganic and organic particles of different sizes. The clay minerals and the decomposed organic matter (humus) make up the very smallest particles (sometimes called colloids). Because of their very small sizes and their chemical structures these two components are often closely associated with each other to give clay humus particles. Their importance to the soil is that they offer negative charge sites on which cations (e.g. Mg, Na, K, Ca etc.) can combine and be exchanged.

CLIFF SLOPE

◄ Penck's slope replacement model ►

CLIMATIC HAZARDS

A whole series of hazards which affect areas of human habitation such as flood, tornadoes, drought, fog, lead pollution, frost, snow, blizzard and hail.

CLIMATIC REGIONS

There are 13 major climatic regions in the world whose distribution is the product of the analysis of weather patterns over the years, see Fig. C.7.

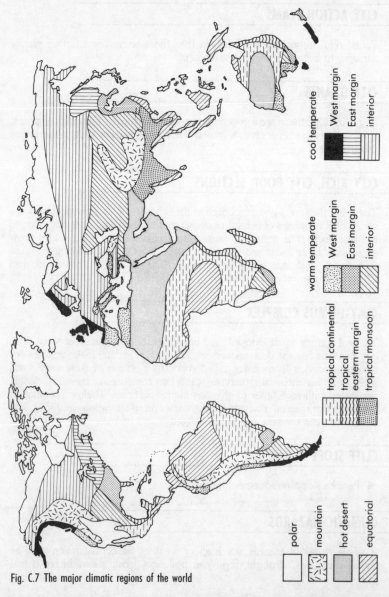

cool temperate
West margin
East margin
interior

warm temperate
West margin
East margin
interior

tropical continental
tropical eastern margin
tropical monsoon

polar
mountain
hot desert
equatorial

Fig. C.7 The major climatic regions of the world

CLINT

The upstanding, flat-topped blocks of carboniferous limestone separated by grykes which go to make up a limestone pavement in those areas where the surface has been glacially smoothed. In areas where there has not been glacial smoothing the surface is much more uneven. Chemical weathering i.e. solution in the grykes, gradually reduces the dimensions of the clint.

CLUB OF ROME

◀ Limits to Growth Report ▶

COAST

The zone between the land and the sea which contains those areas where marine erosion and deposition occur. It includes wetlands where the shallow marshland landscape is either temporarily or permanently flooded.

COASTAL SAND DUNES

Mounds of wind-blown sand which are formed from sand transported from the beach to above the high water mark and accumulated around obstacles. They depend on a constant supply of sand which accumulates around vegetation which traps the sand. However, these dunes are liable to attack from storm waves. At the coastal edge the dunes are small embryo and fore dunes, these are succeeded by the mobile dunes that grow by accumulation and move inland. The final stage is the stabilised dune which is more vegetated than the earlier dunes in the succession. Frequently the dunes are separated by slacks, hollows which extend to the water table and are then filled with small water bodies and marshland wetland communities. It is on the dune landscape that the psammosere plant succession develops. Dunes are threatened by blow outs; storm damage by wind which will remove a large section from the centre of a dune. Such destruction of the dune may be caused by human activity which has reduced the vegetation cover during the summer months. To combat the effects of people on both the vegetation and the morphology a dune landscape and ecosystem needs careful management. East Head at the entrance to Chichester Harbour is one of many dune landscapes where such environmental management has become essential to preserve the unique environment.

COLLECTIVE FARMING

A socialist farming system which is distinguished by its organisation and not its outputs which will be the same as any other farm in similar environmental conditions. The land is state owned and leased to a group of farm workers who then share the proceeds. Some schemes also give the workers land for their own use. It is an agricultural system found in E. Europe, the USSR and China although events of 1989–90 in E. Europe may result in the system's demise. Collectives may also be found in Israel where they are known as **kibbutzim**.

COLONIALISM

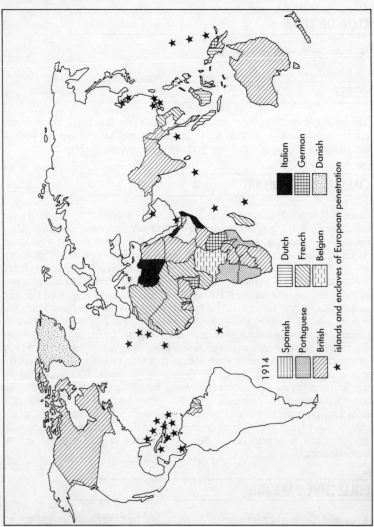

Fig. C.8 European empires around 1914

The rule over people of the **developing world** by the European powers especially during the nineteenth century, see Fig. C.8. The development of much of Africa, South Asia and the West Indies was affected by colonial rule. The colonies were **dependent** on the colonial powers who exploited their mineral resources (e.g. Malayan tin) and agricultural output (e.g. West Indian sugar cane) and often introduced **plantations** to increase the output of products needed by the colonial powers e.g. Malayan rubber. The native population in some areas were used as slave labour (e.g. Barbados) or as cheap labour to work the plantations. In some areas labour was imported from other colonies (e.g. Indians into Malaya). People were often moved forcibly from their land so that white settlers could farm the land or exploit the minerals. **Monoculture** was very common but, on independence, this proved to be a very dangerous foundation for the economy because of price fluctuations, demand volatility and the development of substitute crops elsewhere e.g. beet for cane sugar. On independence, too many former colonies were dependent on too narrow an economic base and too little diversity of export earnings. Colonialism has often continued after independence in the form of **neo-colonialism**.

COMMERCIAL GRAIN FARMING

A form of **extensive agriculture** found in the USA and Canada, the Ukraine and parts of Argentina, Uruguay and Australia where the large-scale production of cereal grains for export (not USSR) has been developed.

COMMERCIAL TREE CROPS

Crops based on the systematic cultivation of a planted forest of a particular species whose sap (rubber) or fruit (oil palm) produce a valuable commercial export resource.

COMMODITY AGREEMENT

International agreements on the production and sale of products such as sugar have been developed to enable cane sugar producers, such as Mauritius and Barbados, to have a continuing access to their former markets in Western European where beet sugar is produced more cheaply. In this way the economy of the producer is maintained because the developed countries agree to buy a given output from the producer countries.

COMMON AGRICULTURAL POLICY (CAP)

The policy of the **European Community** to increase agricultural productivity by promoting technical progress and the best use of labour; to provide a better standard of living for all working in agriculture; to stabilise markets and to provide food at a reasonable price for the consumer. It was established by the Treaty of Rome 1957 and absorbs up to 70% of the EC budget. Its main policy is embodied in the operation of the European Agricultural Guidance and Guarantee Fund (EAGGF) which through price support or **intervention**, maintains rural employment. The guidance side of the fund provides grants for farm modernisation and altering the pattern of landholdings through the process of **land consolidation**. Money is also available for land drainage, irrigation, flood protection and integrated rural development programmes to aid both agricultural and non-agricultural aspects of rural development. **Set aside policy** is a more recent arm of the policy.

COMMON MARKET

◄ European Community ►

COMMUNICATION

The movement of information from one person or place to another either by face to face contact or by the use of **telecommunications**. In the plural the term is synonomous with transportation. The requirements for speed of transfer of information and the nature of the information – i.e. whether it is merely an order, a routine or a discussion of ideas – will determine the location of the establishments where the transfer takes place. For this reason communications technologies have an important bearing on the location of **offices**.

COMMUNISM

A political ideology, based on the writings of Karl Marx, which would evolve from the class war which would be won by the working classes. The victory would result in a classless society with the means of production and distribution owned by the people or proletariat. It is commonly seen as the system of economic control in East Europe, the USSR and China and parts of South East Asia. However, Eastern Europe has never been a truly communist society and could only be seen as a socialist society which was progressing along the path *towards* a communist society. The 1989–90 uprisings there have altered that path. Communism is more strictly adhered to in the Far East (e.g. Vietnam, China) where the interpretation of Marx has been more rigourous.

COMMUNITY

1   A group of people found in a place who may be bound together by real ties such as in a monastery or perceived to be bound together in the view of

the people in the community or in the view of outsiders. It is a term which is used by the property development industry to increase the desirability of new developments. Community, when it does develop, has evolved over time and depends on successive generations being present to assist in the full cycle of social and cultural events of an area. Community studies were a favoured form of Amercian sociological study between 1919 and 1939 and were repeated in British rural areas in the 1950s and 1960s. Wilmott and Young's study *Family and Class in a London Suburb* represents the high point of community studies in post war urban studies.

2   A group of plants and animals which compete for space, complement one another and depend on each others' presence.

COMMUTER VILLAGES

Villages whose growth was attributed to the immigration of commuters. Pahl noted that they grew because of the **decentralisation** of industry and services to new towns, greater affluence, improved public transport and **green belt** policy restricting the growth of some villages so making them desirable. As a result the social composition of the village changed with the newcomers segregated into the newer residential areas. The traditional rural population often found it difficult to compete with the newcomers in the housing market. As a result some counties concentrated housing for the lower paid in certain villages. Low order goods provision in the commuter village declined as more shopped in the towns; the village shop was replaced by higher order provision such as antique shops and restaurants. Public transport provision was also affected since the higher proportion of car owners reduced demand for it. **Counter-urbanisation** is a continuation of the process of movement to rural areas.

COMMUTING

The daily migration of people to and from work. It is normally a short distance, short time movement although around major cities the distances will be larger (up to 150km or more) and the time taken may also be up to 3 hours per day. Due to the time and distance constraints in some cities, there has emerged a body of long-distance commuters who journey to and from their home only at the beginning and end of the work. These long-distance commuters live in an urban **second home**.

COMPACTION

The destruction of the natural structure of the soil and the formation of a continuous dense surface layer of a higher bulk density with a low inflitration rate. Compaction can be caused by natural processes (raindrop impact) or by mechanical influences, e.g. tractor wheels.

COMPACTS

Arrangements between local educational authorities and enterprise agencies to develop the skills of the teenage population through various links between business and the education system. It is another **targetted** form of urban aid aimed at areas such as the East End of London.

COMPARISON GOODS

Those goods sold in shops which depend on the customer making comparisons between competing shops or competing products. Electrical goods and clothing are the main types of comparison good. Shops selling such goods normally cluster within centres to enable customers to make comparisons and to gain from the trade which results.

COMPREHENSIVE REDEVELOPMENT

The complete clearance of a site in an urban area in order to rebuild, often on a completely new street plan. It was the favoured method of improving run-down urban areas in the 1960s although the housing which was built in such areas has met with much criticism since such schemes often meet with public opposition because of the destruction of known environments. Not all schemes are for housing and some of the more recent examples e.g. Les Halles, Paris, and Canary Wharf, London, are for uses other than residential.

CONCAVE SLOPE

◀ Slope form ▶

CONCENTRIC ZONE MODEL

◀ Burgess model ▶

CONGELITURBATION

A general term which refers to the action of frost on the ground surface which results in **frost heaving** and **frost thrusting** and **patterned ground**.

CONNECTIVITY

The degree to which edges are connected to one another in **network theory**. It is measured by the **cyclomatic number**, the **beta index** and the **alpha index**.

CONSERVATION

The preservation and maintenance of buildings or assemblages of buildings. It is normally undertaken to try to retain the best buildings of a period which merit protection. Conservation initially concentrated upon obvious, famous structures although attention has shifted to the less outstanding but nonetheless important buildings from other periods of urban history including 1919–35. Very often conservation is arranged in conservation areas which are areas or assemblages of buildings whose contribution to the beauty of the townscape is the completeness and unity of the ensemble. Building conservation was given statutory backing by the Civic Amenities Act 1967. The buildings which are conserved are known as *listed buildings*.

CONSERVATION (RESOURCE)

A method of ensuring the continued availability of a **non-renewable resource** by making more efficient use of it by **recycling** and **substitution**. Conservation may also mean improving production methods so that less of the depleting resource is used e.g. thinner coatings of tin. It may also involve energy conservation either in the production process or by recycling so that further new materials are not needed e.g. recycling uses less energy than the production of new glass.

CONSERVATIVE MARGINS

◀ Plate tectonics ▶

CONSTANT SLOPE

◀ Slope form ▶

CONSTRUCTIVE MARGINS

Also known as extrusion zones.
◀ Plate tectonics ▶

CONSTRUCTIVE POPULATION PYRAMID

◀ Age-sex pyramid ▶

CONSUMER SERVICES

That part of the **tertiary sector** which is involved in the selling and retailing of products.

CONTAINERISATION

A means of handling cargo which was developed in the 1960s and which reduces the handling costs. The containers can be moved by road and rail and transferred to ships. The standardisation of the size of the containers improves efficiency although they do require the construction of specialist handling areas and cranes.

CONTINENTAL DRIFT

The name given to the theories of **plate tectonics** developed by Wegener in 1915.

CONTINUOUS RESOURCES

◀ Reneweable resources ▶

CONTOUR PLOUGHING

The agricultural practice of ploughing along the natural contours of the land rather than up and down slope. It is thought that such a practice might help in the conservation of soil against removal by water erosion. It also assists the **infiltration** of precipitation and slows or prevents **sheetwash**.

CONTROL POLICY

◀ Population policy ▶

CONVENIENCE GOODS

A term used in retailing to define those goods such as groceries which are needed for everyday life and are bought frequently. It is also used for goods which are easy to use and which have replaced more labour intensive methods of domestic activity, e.g. spray polish rather than tins, pre-prepared food and toasters.

CONVEX SLOPE

◄ Slope form ►

COOL TEMPERATE EAST MARGIN CLIMATE

The climate is one where precipitation is fairly constant throughout the year associated with **depressions** whose strength is accentuated in winter by the contrast between the cold continental interior and the less cold ocean. The climate of the New York area is an example. Temperatures are much more extreme than in the west coast temperate climate with the effects of the continental winter and the easterly movement of air lowering temperatures to an average of below zero for three months of the year. In summer the temperatures can become very high as tropical air pushes north warmed by the ocean currents.

COOL TEMPERATE INTERIOR CLIMATE

The climate associated with Moscow, although the further east one travels the more extreme the climate. Precipitation occurs all year round but with a summer maximum from frontal rainfall and thunderstorms. In winter much of the precipitation may fall as snow. The temperature pattern is markedly seasonal ranging from well below zero to above 20°C in the summer months.

COOL TEMPERATE WESTERN MARGIN CLIMATE

A complex climate based upon the interaction of the global circulation, the movement of **depressions**, the pattern of land and sea and ocean temperatures. The main features are a high pressure which lies offshore and towards the tropics which is composed of tropical maritime air and a low pressure which lies polewards (over Iceland in Europe) around which the depressions track. In winter a polar continental high pressure is often found to the east and this may spread toward the western margins. The main air mass which inflences the climate is polar maritime which approaches the western margins over ocean and is modified by this in different ways at different seasons. In the warm sector of depressions it will have been warmed. The passage of the depressions across the region will be partly controlled by the position of the **jet stream** and its relationship to the **Rossby wave**. The variability of the weather in the zone is caused by the variety of air masses which may come to affect the region. In summer a **blocking anticyclone** from the Azores may deflect depressions to the north and lead to the drought conditions of 1976 and 1989. Similarly, in winter the polar continental air will move east from central Europe and also push the depression track north or south resulting in cold snaps. However when the depression track is across the region then winter will see the characteristic succession of depressions which have brought so many storms to southern England in January and February 1990. In summer, the passage of the depressions will bring unsettled weather and a generally wet summer as in 1988.

COOMBES

A term sometimes used to denote the shorter **dry valleys** which are found on the scarp face of the chalk in southern England.

CO-OPERATIVE FARMING

A voluntary grouping of farmers to gain cheap inputs of seeds and fertilisers by buying in bulk and to profit from economies of scale output from sharing harvesting costs or processing costs. Much of European wine production comes from co-operatives. They are found in the **developing world** where they are used to aid the development of agricultural production through the common purchase and use of machinery and the sharing of marketing costs.

COPPICING

The cutting back of trees in the deciduous forest to the stump to produce a series of shoots which grow rapidly. It was normally used on hazel and sweet chestnut to provide fuel, and areas of coppice were systematically cropped every 10–15 years. Very often larger oak trees provided a cover to draw the shoots up towards the light and this was known as coppice with standards. The practice increased the yield of the forest but also adapted the ecosystem.

CORAL HALOSERE

The reef ecosystems may be fringing coral, attached to the shore or separated from it by a lagoon, barrier coral, offshore acumulations running parallel to the coast and often lacking any channel to the sea beyond from the lagoon, and atolls, circular shaped coral areas mainly found in the Pacific Ocean, possibly built up on a volcanic core as sea level rose after the glacial period. This is a tropical ecosystem found where the sea temperature is 21°C and above and where there is some form of platform on which the coral may grow in the clear, silt-free water. It is more common on the east coast of continents. The animals which give rise to the reef are carnivorous, feeding on zooplankton, and build an external skeleton which gives the reef its stability. Reefs are threatened with destruction by natural predators such as the starfish attacking the Great Barrier Reef, destruction to provide navigable channels and removal for souvenirs.

CORE FRAME CONCEPT

This develops the concept of the **CBD** to take account of a city's dynamism. The core is the CBD whereas the frame contains the land uses normally associated with the fringes of the CBD. Transport termini, warehousing, wholesaling, some manufacturing, automobile sales and servicing, some residential uses (mainly lower socio-economic groups) and some special professional services are associated with the frame, as shown in Fig. C.9.

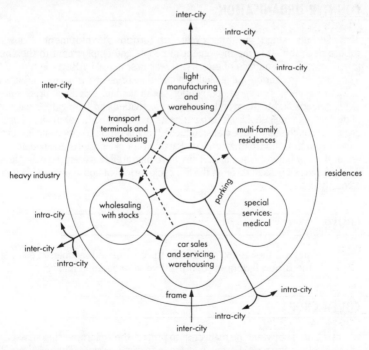

Fig. C.9 The core-frame concept ⟶ primary goods flow
- - -► secondary goods flow

Parking lots will also be found here. The core will be **assimilating** areas such as those with professional services and where **gentrification** is in progress while other areas of poor quality retailing e.g. cheap second hand goods are being **discarded** by the core.

CORRASION

The process of mechanical erosion of a surface by the material being transported over it which results in the **abrasion** of the surface.

CORRELATION

The degree of relationship between two variables which may be plotted graphically or tested statistically by one of a series of tests such as Spearman's rank correlation.
◀ Dependent variable ▶

COUNTER-URBANISATION

The second stage in the **cycle of urban development** following **suburbanisation**. It is the movement of people and employment to the small towns and villages outside of the city. It was first noted by Berry in the USA. The causes are varied but include; population pressure in cities, affluence, modern means of transport, perceptions of urban life, locational preferences of households and tertiary and quaternary activities and new career patterns of dual income households. It also suggests that the traditional model of urban growth through suburban extension fuelled by migration to cities and life cycle moves to outer suburbs has ended. Partly in its place there are people who are of the city but not living in it. They have migrated between the outer areas of cities and not passed through the **life cycle model** of house ownership in the same way.

COVER

The percentage of the ground overlain by the crowns and canopies of plants. It is used in the **Braun-Blanquet rating system** of vegetation mapping.

COVER CROP

A crop planted between the main crop to protect the soil from erosion, used to greatest effect in tropical areas subject to torrential rains. The crop may be ploughed in to provide further nutrients. In orchards grazing land provides a cover crop for sheep.

CRAG AND TAIL

A feature of glaciated areas where a resistant rock has withstood erosion and has protected the area in its lee from erosion to form the tail. In some cases the tail is formed of till. Edinburgh Castle is on the crag while the Royal Mile is on the tail. Stirling Castle is sited on another crag and tail formation which provides a good defensive site.

CREEP

The slow movement downslope of material under the influence of gravity which may occur on any weathered surface. Movement is slow and imperceptible although it is faster nearer the surface. The angle of slope will influence the rate of creep. It is a form of **mass movement** which is most common in humid temperature climates and **periglacial** areas where it is a form of **solifluction** called frost creep. Small terracettes and the non-vertical appearance of walls and vegetation are products of the process.

CREVASSE

A crack or fissure on a glacier which extends down into the glacier. It provides a passage for surface meltwater to move into the glacier as an **englacial** stream often filling the crevasse with debris which, when deposited, becomes a **kame**.

CROSS- SECTION AREA

◀ Channel variables ▶

CRUDE BIRTH RATE

The number of babies born per year per thousand population. It is crude because the population figure includes those who cannot give birth to children i.e. men, children and the elderly. A refinement of this measure is the **age specific birth rate** which is calculated for each five year age group or for women of child bearing age, normally 15–45 years old. It is the total births to women in these age bands divided by the number of women in each age band times 1000.

In 1985 the crude birth rate in Nigeria was 48/1000, Bangladesh 45/1000, Brazil 31/1000, United Kingdom 13/1000 and Sweden 11/1000. Birth rate is influenced by; i) the number of women of childbearing age, ii) the age of marriage, iii) the number of married women, iv) society's norms for family size, v) the availability and acceptance of methods of family planning, vi) levels of education particularly of women and vii) economic conditions in the country. Birth rates are the most commonly used indicator of fertility.

CRUDE DEATH RATE

The total number of deaths per year per thousand population. It is crude because it includes those whose chances of death are very small such as the young. A refined method is the **age specific death rate** which is the number of deaths per age band per thousand members of that age band. In Nigeria the crude rate in 1985 was 17/1000, Bangladesh 17/1000, United Kingdom 12/1000 and Sweden 11/1000. The rate is influenced by age, sex (males die younger than females), social class, occupation and life-style.

CULTURAL EUTROPHICATION

Accelerated **eutrophication** caused by human action. The sources of the aquatic pollution are domestic sewage and especially detergents, urban run-off which includes animal waste, garden fertilisers and oil products, industrial wastes, nitrates from poorly managed farmland which has received an overapplication of fertilisers, and slurry from animal waste draining directly into a stream. The consequences are the gradual death of a lake system with only fungi and bacteria remaining, hindered shoreline development and even declining property values and a smell/taste to the water which reduces the recreational utility of the water body.

CULTURAL GEOGRAPHY

The study of communities and societies and their relationship to their environment. It involves a series of themes which link people to their environments although more modern aspects of the specialism involve the study of the attributes of culture such as language and religion.

CULTURAL RESOURCES

◄ Capital resources ►

CULTURE OF POVERTY

The lifestyle in areas of poverty which helps to perpetuate the **cycle of poverty**. It is a theory which is espoused by those who see one of the causes of the inner city problem as the way of life in those areas which has to be changed so that the forces which promote that culture are broken.

CYCLE OF POVERTY

Poverty is self-perpetuating and is transmitted from one generation to the next. Children from poor homes start with a disadvantage, receive little support at home, underqualify at school, have problems finding work and so remain poor. Poverty, ignorance and violence are passed on to each generation. Many urban programmes are based on the hypothesis that it is possible to break into this cycle and combat poverty but this ignores the political structure in which that poverty exists.

CYCLE OF RESOURCE EXPLOITATION

rate of discovery of a resource

rate of production of a resource

rate of increase of proven reserves

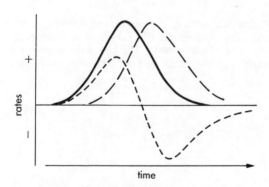

Fig. C.10 Cycle of exploitation of a resource

A four-stage model of resource exploitation shown in Fig. C.10 commencing with exploration and discovery when reserves are greatest, followed by exploitation of the major find and continuing increase of reserves with further discoveries, exploitation of the minor finds as reserves decline, and exhaustion of the finds with increasing recycling and substitution.

CYCLE OF URBAN DEVELOPMENT

A sequence of urban growth and change in terms of population movement as shown in Fig. C.11.

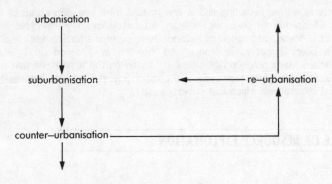

Fig. C.11 Cycle of urban development

CYCLOMATIC NUMBER (u)

$u = e - v + s$

The cyclomatic number equals the number of edges (e) minus the number of vertices (v) and plus the number of sub-graphs (the subsidiary or unconnected graphs). It is a simple measure of **connectivity**. The higher the number the more complete the connections between the vertices and the more complex network.

DAIRY FARMING

An agricultural system which specialises in the production of milk and other dairy products. It is a form of **intensive agriculture** because the labour and capital inputs are high. The system is mainly focused upon temperate areas.

DATA BANK

A storage system for data within a computer. These banks have become increasingly large as computer power has increased and contain many million entries of data. Geographical information systems (GIS) depend on the availability of data from a wide variety of data banks.

DAVISIAN SLOPE PROFILE DECLINE

A model of slope evolution in which a balance between weathering and transport results in a declining angle of slope. On the lower slope, transport has to remove all the material weathered at that point besides all the material removed from above and so weathering of the lower slopes is progressively slower as the angle of slope declines. Initially slope decline on the upper slope is faster but it slows as the angle declines because of the difficulties of transporting the material away from the point of weathering. It is an explanation which best fits the older phases of landscape evolution.

DEATH RATE

◄ Crude death rate ►

DEBRIS SLOPE

◄ Penck's slope replacement model ►

DECAY CHAIN

Dead plant and animal matter is fed upon by a range of **decomposers** which return the nutrients to the soil. Because there is no fresh input of nutrients to the system recycling must take place and the decayed matter is ready for recycling into the system.

DECENTRALISATION

The movement out from the central districts of a city to the suburbs of a city or beyond into the city region of commercial and industrial fuctions. Sometimes, in the case of major cities, it can be to other parts of the national urban system. Industry began to decentralise to suburban sites after 1945 besides moving to the **development areas**. Decentralisation to **new towns** was encouraged. More recently office functions have followed a similar pattern of movement to suburban centres e.g. Croydon, to development regions e.g. Swansea, and to other towns in the south-east, e.g. Eagle Star Insurance to Cheltenham. Wholesaling has also decentralised from city centre sites where there are pressures to use the land for commerce, retailing and tourist functions, to other sites in the city e.g. Covent Garden to Nine Elms, or to the outer suburbs e.g. Les Halles to Rungis in southern Paris. Retailing is also decentralising to new covered shopping centres in the outer city or on the rural–urban fringe and to retail warehouse parks such as those occupied by many DIY stores.

DECOMPOSITION

All the chemical/physical and biological processes involved in the decay and conversion of material in organic form to an inorganic state.

DEFLATION

A dominant process in arid and semi-arid regions by which wind removes sand, silt and light clay from the ground surface. It can also occur in more humid areas particularly after a dry spell when strong winds will remove topsoil from ploughed fields. It is common in East Anglia and on the large unhedged fields of much of northern Europe especially in the winter months when fields await the germination of autumn planted cereal crops. Deflation is also responsible for the movement of sand in dunes. Deflation was the process which carried **loess** from the ice sheets to the areas of deposition further south.

DEFORESTATION

The complete felling and clearance of an area of forest for alternative land uses. Forest clearance was an important activity in Great Britain in the Middle Ages. Today forest clearance especially in the Amazon basin is a cause for concern because of the role that the forest plays in the global ecosystem. Deforestation which is not carefully managed leads to **soil erosion** and can lead to the loss of species.

DEGLOMERATION ECONOMIES

The negative operation of the forces of **agglomeration** on the productive units in an area which occurs once the benefits of agglomeration become costs due to transport congestion, high wage rates and loss of amenity. Deconcentration and **decentralisation** may follow from the realisation that deglomeration forces have exceeded the benefits of agglomeration.

DEINDUSTRIALISATION

The decline in the contribution of manufacturing industry to the economy of a region or a country. It is manifest as a decline in the number and proportion of manufacturing jobs. It is caused by cheaper products being manufactured elsewhere and imported, products being at the end of their **product life cycle**, production being shifted to low cost locations outside of the region or the country, and taxation levels forcing migration. It results in job loss and often there are policies to counteract these losses but the new jobs often involve different skills and are more suited to female labour than the male dominated industries which are affected by deindustrialisation.

DELTA

A fan-shaped area of deposition at the mouth of a river into the sea or a lake. As the stream enters the sea the heaviest load is deposited and the lighter material is only deposited in the more saline waters. The delta grows if the rate of supply of sediment exceeds the rate at which the processes of coastal erosion remove the material. The Nile delta which has the most common form, is no longer growing because the Aswan High Dam has cut off much of the supply of sediment which enabled it to grow. The Mississippi delta is more extended into a **birdsfoot** shape due to the rapid supply of sediments from its vast catchment.

DEMOGRAPHIC CHARACTERISTICS

The characteristics of a population expressed in terms of its **population structure**, age, sex, marital status, fertility and natural change. Ethnic composition, rates of change and socio-economic characteristics of populations can also be studied.

DEMOGRAPHIC TRANSITION MODEL

A model of the general changes through time of the population of a country consequent upon the interaction between **birth rate** and **death rate**. The model is based on the experience of countries in the **developed world** and can be linked to the overall processes of development. Therefore, the model has a series of stages reflecting the demographic and economic development of a country; Stage 1 high stationary; Stage 2 early expanding; Stage 3 late expanding and Stage 4 low stationary; see Fig. D.1. Some researchers do

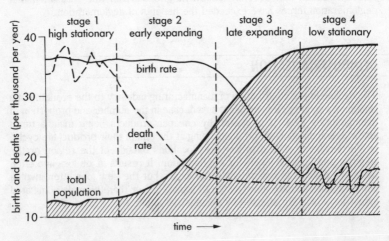

Fig. D.1 The stages of demographic transition

suggest that there is a fifth stage, late declining, which can be found in countries where the death rate is almost continually above the birth rate leading to natural decrease of the total population. The model has been criticised because it is not necessarily true that all countries will progress through the same demographic and economic stages.

DEMOGRAPHY

The empirical, mathematical and statistical study of population. **Population geography** is the branch of geography which has as its focus demographic processes. The data used in demography is normally obtained from national censuses which are held at least once every decade. Other surveys can also provide data. Some countries no longer have censuses and use alternative sources of data such as registration data. This is the case in Denmark.

DENATIONALISATION

◄ Privatisation ►

DENUDATION

A broad term used to describe all the processes which wear down the earth's surface. It refers to all the processes which start with **weathering** and includes **abrasion**, **corrasion** and **erosion**. **Transport** of material is an essential part of denudation because that exposes further material to the processes of denudation. The rate of denudation varies between areas according to the climate, the nature of the landscape and the lithology.

DEPENDENCE

◄ Frank's dependency theory ►

DEPENDENCY RATIO

The percentage of *economically non-active population* divided by the percentage *of economically active population*. It can be subdivided into the young **dependent population**(i.e. those under the age of 15 years and the elderly dependent population (i.e. those over 60). There are complications here because of differential retirement ages for men and women and the fact that many continue working beyond retirement age e.g. Mrs Thatcher as Prime Minister! A high ratio indicates either a country whose population is expanding which is Stage 2 of the **demographic transition,** or a country in Stage 4 whose population is ageing due to a low rate of **natural increase.**

DEPENDENT POPULATION

◄ Dependency ratio ►

DEPENDENT VARIABLE

It is the y, abscissa or vertical axis on a graph as opposed to the **independent variable** on the x, ordinate or horizontal axis which is plotting the relationship

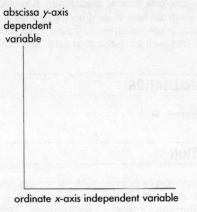

Fig. D.2 Labelling axes

between two variables, see Fig. D.2. The plotting of data in this way is an essential prerequisite for **correlation** tests.

DEPOSITION

The final stage of **denudation** when the energy to transport sediments declines and landforms of deposition result. Deposition is the final stage of all processes of erosion and deposited sediments may be reworked by different processes, e.g. wind borne **loess** and ground **moraine** will be reworked by the processes of river erosion. It may be called *aggradation*.

DEPRESSED AREAS

◀ Development areas ▶

DEPRESSIONS

Depressions are areas in the atmosphere in which the pressure is lower than in the surrounding areas, the central pressure varying from as low as 930mb to about 1020mb. The term is sometimes used to describe weak tropical cyclones, and is also applied to relatively small areas of low pressure generated by orography or by the dynamics of polar air flow. However, the most common usage of the term depression applies to the extra-tropical cyclone or mid-latitude depression.

Typically such depressions are associated with fronts separating warmer, moister tropical air from cooler, drier polar air. Commonly three stages can be identified. An initial wave depression where a moving wave is apparent on the front separating the Tm and Pm air masses. One arm of this wave is a warm front and the trailing arm a cold front. This wave depression may move rapidly and deepen, developing into a full warm sector depression. The final stage is when the steeper and faster-moving cold front catches up with the warm front forcing the Tm air aloft. This forms an occluded front and in the final stage the bulk of the warm front will have become occluded. It is worth noting that occluded fronts are not inactive features, the forced rise of the Tm air sometimes giving very heavy precipitation. The three stages typically last four days but can last considerably longer.

Mid-latitude depressions transport warm air towards the poles and are part of the general circulation of the atmosphere which attempts to equalise out the gain of radiation at low latitudes with the net loss in high latitudes. However, it is not just the contrast between the warm and cold air which controls the development of such depressions. An important feature of mid-latitude depressions is their relationship to the polar front jet stream. The waves on that jet stream create the necessary zones of upper level divergence which enable the surface depression to deepen (greater upper level divergence than low level convergence leads to a fall in pressure).

DESERT BIOME

The vegetation pattern which is associated with the world's hot desert regions although one must acknowledge the existence of other desert types. Most have a lack of permanent vegetation cover due to the aridity, and the plants which do grow are adjusted to the aridity to survive in **ecological niches** such as **oases**. Soils are skeletal and often saline. Plants are ephemeral in that they complete their life cycle after one brief rainstorm and leave seeds to lie dormant until the next precipitation. Drought resistant plants are called xerophytes and are adapted to the aridity with tap roots, branching roots, small, pale leaves and an ability to store water in their tissue. On the borders of the deserts the biome becomes a semi-arid biome which resembles the **tropical savanna** or the **mediterranean biomes**. It is in these areas that the process of **desertification** is at its strongest. The biome has been least altered by human adaptation although **irrigation** has been a traditional method of adaptation either by traditional means or by modern drip feed systems characteristic of those states which have vast oil revenues to spend on making the desert bloom. **Salinisation** is a problem with such irrigation schemes.

DESERTIFICATION

The extension of desert conditions into areas where they did not exist previously, see Fig. D.3. The cause is partly climatic, to do with the **Inter-Tropical Convergence Zone** (ITCZ) not following its normal pattern of

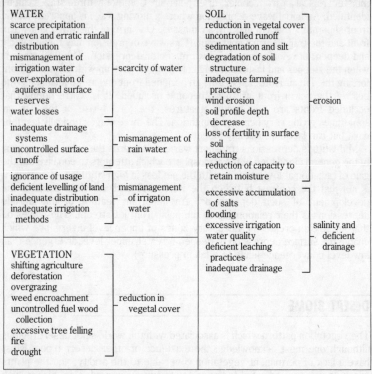

Fig. D.3 Physical and biological processes causing desertification

movement to bring summer rain to an area and partly to the action of people through pressure of numbers, pressure to grow cash crops, overgrazing and clearance of timber from the land.

DESERT PAVEMENT

◄ Stone pavement ►

DESERT VARNISH

A form of crust found in deserts which is made up of a layer of iron oxide or manganese oxide drawn up from the rock by **capillary** action. In the Namib desert it is formed of gypcretes and in Lake Eyre, Australia of silcretes. In all cases the crust becomes polished.

DESIGNATED DISTRICTS

These were established by the Inner Urban Areas Act and were the lowest tier of support for urban regeneration after **partnership authorities** and **programme authorities**. Fifteen areas, e.g. Newham and Barnsley, received support.

DESTRUCTIVE MARGINS

◀ Plate tectonics ▶

DEVELOPED WORLD

Sometimes called the **First World** and 'The North' by the **Brandt Report**, the developed world is formed of the countries of the economically advanced world whose wealth is based on a long period of development. Other states became developed more recently but these are few in number.

A further term for these states is More Developed Country (MDC). It is a term which is synonymous with the capitalist states where the ownership of capital is in private hands and the economy is driven by the quest for profit.

DEVELOPING WORLD

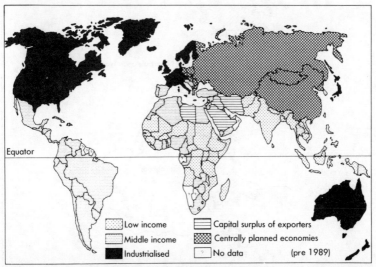

Fig. D.4 The World Bank groupings of countries, 1981

Sometimes called the **Third World**, 'The South' by **Brandt** and **Less Developed Country** (LDC), the developing world is formed of those poor undeveloped economies in Africa, Latin America and Asia which are still in the early stages of economic, technological and social development, see Fig. D.4. The **Newly Industrialising Countries** (NICs) are also part of the developing world.

DEVELOPMENT AGENCIES

The Scottish Development Agency (SDA) and the Welsh Development Agency (WDA) are government-funded bodies attracting industry to these regions by industrial investment and promotion. Their activities are closely prescribed.

DEVELOPMENT AREA

Areas scheduled to receive aid from the government because of their poor performance, see Figs. D.5a) and b). They were given assistance which was

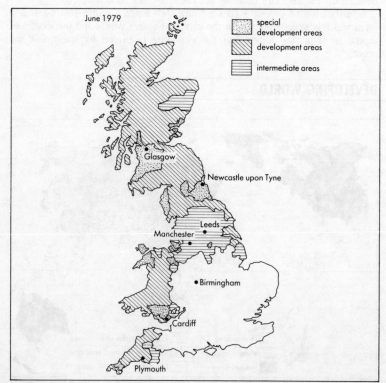

Fig. D.5a)

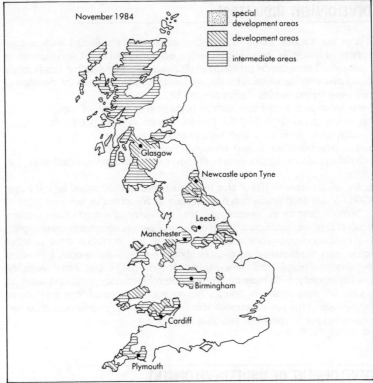

November 1984

special development areas

development areas

intermediate areas

Glasgow

Newcastle upon Tyne

Leeds

Manchester

Birmingham

Cardiff

Plymouth

Fig. D.5b) Rolling back the map of regional aid

to enable work to be taken to the workers and have been called **depressed areas** and **special areas** at various times in the past. Besides the 'carrot' to draw jobs to the regions there was a 'stick' in the form of controls on the growth of industry especially in the South-East. Between 1947 and 1984 **industrial development certificates** were needed for industrial building in the controlled areas. Similarly **office development permits** were required between 1964 and 1979. In 1987 the government replaced the mandatory assistance to all firms locating in these regions with a **discretionary** policy of helping firms who would not have gone to an area without assistance. However, the 1984 map of development areas has been retained specifically to enable firms to qualify for assistance from the **European Regional Development Fund**.

DEVELOPMENT CONTROL PROCESS

◀ Town planning ▶

DEVELOPMENT INDICATORS

These are various forms of statistics which, when collected from a wide variety of countries, provide data on the economic and social development of countries. There can be problems of obtaining common data but this is minor compared with the inequalities that such data reveals between the **developed** and **developing worlds**. Indicators can be

gross showing the total size such as population or volume of exports,

per capita, gross divided by the population such as GNP per capita

density, gross divided by area such as population density

ratios, population per subject area, e.g. people per doctor,

percentages of total figures such as the percentage who can read and write, i.e. literacy,

index using a base = 100 so that food production in 1980 would be 100 and in 1990 95, i.e. food production had declined by 5% in the decade.

Some indicators in common use are; the share of exports from primary products, labour force data, value of exports per capita, energy consumption per capita, cement production, percentage of the population living in urban areas, cars per thousand people, telephones per thousand people, televisions per thousand people, daily newspapers per thousand people, **infant mortality, life expectancy, birth rate, death rate, natural increase**, population under 15 years old, food intake in calories per capita, an index of food production, percentage of the population with access to safe water, population per doctor, population per hospital bed and adult literacy.

◀ GNP, GDP ▶

DEVELOPMENT OF UNDERDEVELOPMENT

◀ Frank's dependency theory ▶

DEVELOPMENT POLES

◀ Growth pole theory ▶

DEVELOPMENT STAGE MODEL

A model which traces the evolution of the four sectors of the economy, through time as in Fig. D.6. Stage 1 is a self-sufficient **primary sector**, stage 2 is dominated by the primary sector but with specialisms developing within it; stage 3 sees the **secondary sector** develop around a small range of goods; stage 4 sees a dominant secondary sector with a wide range of goods demanded by a wealthy society; stage 5 contains the growth of the **tertiary sector** to trade these products; and stage 6 sees the emergence of a **quaternary sector** to develop further the range of technologies and ideas.

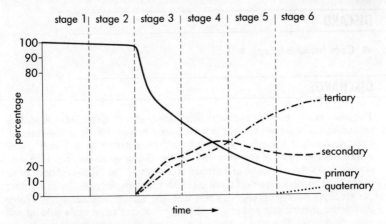

Fig. D.6 The development stage model

DEVELOPMENT STRATEGIES

There are several different strategies which may be adopted to assist in the economic development especially of developing countries. Some of the following may be used at various stages in a country's development; technological transfer, intermediate technology, industrial development with the aim of becoming a newly industrialising country, import substitution, agricultural development schemes and rural development programmes such as the green revolution, improved trade, the development of tourism and the use of aid.

DIFFERENTIAL EROSION

The more rapid rate of erosion in some areas than in others caused by variations in the resistance of the rocks to erosion. Differential erosion can be seen between hard and soft strata and within a single rock type in a beach formation where the waves exploit the zones of weakness of the joint and bedding planes.

DILATATION

The break up of rock caused by pressure release when a weight is removed from it. It often refers to the pressure release found after a glacier has left an area of bare rock which then expands and breaks up. It also causes the appearance of apparent joint planes on tors in granite areas.

DISCARD

◄ Core frame concept ►

DISCHARGE

The rate of flow of a river at a particular point over a given period of time. It is related to the volume of water and its speed or velocity. It is measured in cubic metres per second or cumecs by a flow meter or by a V notch weir where the geometry of the weir enables the discharge to be calculated with precision. As the discharge of a stream increases so the **cross-sectional area** of the river increases to deal more efficiently with the increase by dissipating the stream's energy through friction with the stream channel sides. Water abstraction for drinking and irrigation can reduce discharge while effluent will raise discharge. The rate of discharge can be affected at any one point by culverting, bridges, weirs and reservoirs.

DISCRETIONARY ASSISTANCE

◄ Development areas ►

DISPERSED SETTLEMENT

A pattern of rural settlement where the population lives in scattered dwellings. The degree of dispersion can be measured by **nearest neighbour analysis**.

DISTRIBUTION OF INDUSTRY ACT 1945

The first government initiative to intervene in market forces affecting the distribution of industry. It set up the **development areas**.

DISTRICT PLAN

◄ Town planning ►

DOLINE

A surface depression found in areas of **karst** where intensified **chemical weathering** creates a hollow which may connect down to underground drainage and an **underground cave system**. The doline is known as a shake hole in England.

DRAINAGE BASIN

◀ Catchment basin ▶

DRAINAGE DENSITY

The ratio of the length of all streams in a **catchment basin** to the total area of the basin. It varies with rock type and is highest on impermeable rocks and also varies according to climate and character of the landscape.

DRIP

◀ Throughfall ▶

DRIVE TO MATURITY

◀ Rostow's stages of economic growth model ▶

DROUGHT

The lack of water from precipitation in relation to the normally expected amount. It occurs in the UK in summer when **blocking anticyclones** dominate the weather pattern for a long period of time.

DROWNED BEACHES

The converse of the **raised beach**, a beach which was formed during a period of lower sea-level and submerged once an **eustatic change** caused by the melting ice caps had raised the level of the seas. These beaches can become the source of some of the supplies of **offshore gravel**.

DRUMLIN

A feature of glacial deposition composed of **till** or boulder clay which has an elongated, humpbacked appearance. The shape of the drumlin is thought to be moulded ground **moraine** although others have suggested that the deposition occurred around a rock or frozen drift core. Drumlins are often found in swarms in drumlin fields giving a 'basket of eggs' topography. The oval shape of the drumlin is related to the direction of the origin of the ice which deposited the drift with the long axis being parallel to the direction of ice flow.

DRY VALLEY

A valley found mainly in chalk landscapes but also present in other limestones. Geomorphologists do not agree over the method of formation. The following theories have been used to explain their occurrence:
– a drop in average precipitation which lowers the water table and so leaves valleys dry;
– the recession of the escarpment which would also result in the lowering of the water table and the drying up of streams;
– the presence of water during **periglacial** periods which could then erode the frozen, impermeable rock;
– the effect of river capture leaving a valley dry;
– the disappearance of a stream underground caused by a **swallow hole** opening up where there was not one before, as happens in **karst** scenery.

DYKE

1 An **intrusive volcanic** landform formed when molten magma is forced upwards like a wall cutting through the bedding planes of the country rock. Depending on the resistance to erosion of the dyke and the country rock, it will either form a wall protruding above the surface or a trench in the surface. In some areas dykes are said to swarm, i.e. there is a great concentration of them.
2 An embankment built to prevent marine flooding or river flooding in the Netherlands. In the latter case they are really **artificial levees**.

DYNAMIC EQUILIBRIUM

A descriptive term which may be applied to a climax vegetation in that it is in a state of balance with the forces which are being applied to it. Therefore the plant community neither alters its composition nor changes its location because the same young plants and animals replace each other. **Succession** has halted. It is also applied to other parts of the physical system, air, water or solids.

EARLY TRANSITIONAL SOCIETY

◀ Mobility transition model ▶

EARTHQUAKE

Shock waves produced by sudden movements of the earth's crust along the zones between the earth's plates. This releases **seismic energy** which travels as shock waves through the earth. The study of these waves enables seismologists accurately to pinpoint the **focus** or **epicentre** of the quake. The strength of an earthquake is measured on the **Richter scale**: this scale is **logarithmic** so that a quake of 6.0 is ten times as powerful as one of 5.0.

Earthquakes may occur beneath either the land or sea and in the latter case they give rise to **tsunami** or tidal waves which on reaching land can cause extensive flooding and destruction. Osaka has built special barriers to counteract tsunami flooding. Another danger is that of **liquiefaction** of the rocks caused by the shaking of the ground which then behaves like a jelly, making buildings unstable as happened in the Mexico City quake. Indirect effects include fire, ruptured water and sewage pipes, power cuts and damage to bridges and raised highways, all of which happened in the 1989 San Francisco quake. Landslides, flooding and crop destruction are also widespread. Earthquake monitoring to combat the threat is common and ranges from highly scientific procedures such as the use of laser beams to detect movement on the San Andreas Fault in California to the use of animal indicators in China. Fluid injections into the fault zone are being tried in California to reduce the shocks. Building heights and construction are also strictly controlled.

◀ Plate tectonics ▶

ECOLOGICAL CAPACITY

The capability of the natural environment to sustain itself without damage. It is an important concept in rural recreation. For example, ecological capacity may be threatened by too many visitors to the egg locations of turtles on the beaches of Malaysia. It may also be the capacity of a footpath to a beauty spot to repair itself after it has been pounded by visitors' feet for three months.

ECOLOGICAL NICHE

The ideal location in a habitat for a specific plant or animal. Aspen grow in such niches in the **temperate grassland biome**.

ECOLOGY

The study of the interrelationships between plants, animals and their environment.
◄ Ecosystem ►

ECONOCLIMATE

A term given to the effects of climate on economic activity such as the effects of a lack of snowfall on the economic fortunes of Alpine ski resorts.

ECONOMIC GEOGRAPHY

The study of the distribution of economic activity. It includes the study of the factors and processes influencing the location of the activities and the changes in those locations. It normally includes the study of agriculture, forestry, fishing and mining, **(primary sector)** the study of manufacturing, **(secondary sector)** and the study of the service or **tertiary sector** which includes **transport** and trade.

ECONOMIC RENT

The value of a crop minus the production and transport costs in the **Von Thünen model**. It explains why the net returns from a crop will vary from place to place according to the distance from the market and transport costs. It is also used to explain why agricultural systems vary from region to region.

ECONOMY OF SCALE

The financial gains that can be made from large scale production. The larger the company and its output the greater the savings that can be made from the cost of producing each item. These economies may be gained through internal specialisation by the company or by producing more at each piece of equipment. In retailing the supermarkets rely on economies of scale when purchasing products because the more they buy the less they pay for each unit.

ECOSYSTEM

The interaction between a set of living objects in their environment which can be studied at all levels from that of the global ecosystem to the small scale patch of grass. It is comprised of **autotrophs** which fix light energy and use inorganic materials from the soil and atmosphere and **heterotrophs** which feed upon the autotrophs to redistribute the energy and nutrients through the system, as shown in Fig. E.1.

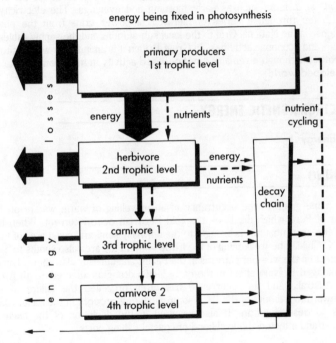

Fig. E.1

EGYPTIAN TYPE

◀ Population-resource ratio ▶

ELECTORAL GEOGRAPHY

A branch of **political geography** which examines the results of elections and the proportion of votes cast for each party. The changes between elections by ward or constituency can be examined in relation to other variables such as age groups, income and even country of origin in some special types of elections, e.g. the foreigner assembly in Cologne.

ELECTRICAL POWER

A secondary source of energy which is derived from primary sources such as coal and oil. It is generated in power stations which are either base load in that they produce *all of the time* or peak load in that they produce for the *main periods of demand* such as early morning and evenings. The electricity is distributed through a series of hierarchical power grids from the power stations via the National Grid to the local sub-stations and the supply cables to homes and economic activities. In Great Britain it is an industry which is to be **privatised** although it remains a **nationalised** activity in most countries in the developed world.

ELECTROMAGNETIC ENERGY

◀ Energy ▶

EL NINO

The name given to the occurrence of an upwelling of warm waters off the coast of Peru which displaces the cold Peruvian **ocean current**. When this happens, torrential rain is brought to the normally semi-arid coast area of Peru and N. Chile. The weakening of the **trade wind** in the area also seems to have an effect on the weather patterns in the Pacific and perhaps elsewhere. During a prolonged El Nino effect in the early 1980s droughts affected South Africa and Australia and its occurrence has often preceded the failure of the northward movement of the **ITCZ** so precipitating **drought** in the Sahel which leads to **desertification**. It also resulted in the depletion of the anchovy harvest and affected the livelihood of coastal fishing ports.

ELUVIATION

The removal of soil material in solution or suspension from layer or layers in the soil. Produces eluvial horizons.

EMIGRATION

The movement of people out of an area which may result in population decline. On the whole it is a reaction to **overpopulation** and population decline is rarely the case. Absolute decline did follow emigration after the Irish potato famines of the 1840s. Emigration does result in a lowering of the birth rate because the more fertile, younger people tend to migrate.

EMPLOYMENT MULTIPLIER

◄ Multiplier effect ►

ENCLOSURE

The process of consolidating rural properties, abolition of common rights and the fencing and hedging of fields which either took place by informal agreement or by parliamentary action and produced the English agricultural landscape of the past 100 years. It is a type of land reform.

ENDOREIC STREAM

A stream which is a part of an inward flowing drainage pattern such as the ephemeral streams which flow into Lake Eyre, Australia. This term is also spelt ENDORHEIC in some sources.

ENERGY

Energy is the lifeblood of society because it drives all human activities. It is the means of providing heat, light and movement. All energy except geothermal energy is derived from the sun and even human energy is from that same source.

Electromagnetic energy is that derived from the sun which has been stored in fossil fuels and which is received daily into the natural system.

Kinetic energy is that lost as heat and friction by running water and waves.

Potential energy is the energy stored in an object before it is released, e.g. before a wave breaks.

Nuclear energy is a form of potential energy contained in the nucleus of an atom. Geothermal energy is contained in the earth's interior and only usable when it comes near to the earth's surface as geysers or hot springs. Energy which is used by people is primary energy although some of the stored energy has to be converted to a usable form in power stations to produce secondary energy or electric power. Final energy is that energy which is used and ignores losses in distribution. It may be converted for other uses, e.g. fertilisers and plastics from oil.

ENERGY LOSS

The losses which occur during the transformation of energy in the form of heat, the distribution losses due to leakage and the inefficiency of the machinery producing or using that energy.

◀ Final energy ▶

ENERGY RESOURCES

These are mainly made up of **fossil fuels**, but firewood, charcoal and biofuels such as animal and crop residue are still the main source of fuel for half of the world's population. Fossil fuels are all **non-renewable resources** as is **uranium**, used in **atomic power**, which is dependent on the occurrence of the mineral. **Solar, wind** and wave **power** are all technologies in their infancy but with future potential because they each depend on a **renewable resource**. **Hydroelectric power** is an expensive renewable energy resource most used in the **developed world** although some major barrages provide for the energy needs of some developing countries. **Tidal power** barrages are few in number because of the capital costs of construction, although the potential sites are numerous. **Geothermal** power is used in some countries where the right tectonic conditions prevail.

ENGLACIAL MORAINE

◀ Moraine ▶

ENTERPRISE BOARDS

Established by local authorities to attract private capital to an area, e.g. West Midlands EB. They are now independent of local authority control and represent a private initiative in regional development.

ENTERPRISE ZONE

Introduced in the Budget of 1980, Enterprise Zones are one strategy for the revival of inner city areas. Twenty four zones e.g. Isle of Dogs, Swansea and Corby have been designated, the last, Inverclyde in 1987. No more are being designated and the programme will wind down in the 1990s. Tax burdens were removed from firms locating in the areas; the most significant of which was exemption from rates for 10 years. Planning was simplified and other concessions to ease the establishment of the firm were implemented. The areas vary in size from 200 acres to 1000 acres in Gateshead. There is a problem over the effect on firms just outside of the boundaries. Many zones seem to have acquired warehouse retailing and storage and not new industry. In other cases industry has merely moved site to gain tax advantages, one example being several newspaper companies moving to the Isle of Dogs. So the value of the policy in terms of job creation is questionable. Costs of the policy are well over £200 million.

The enterprise zone strategy is also being used by the government of China to encourage economic development in the area bordering Hong Kong.

ENUMERATION DISTRICT

The smallest area for the collection of census data in the United Kingdom. It is normally about 200 households and the boundaries of these areas can be seen on maps held by District Councils. The data held for EDs is known as Small Area Statistics and is not published like the data aggregated for parishes, districts and counties although it can be obtained as computer print-outs. Some of the data for EDs is based on a 10% sample of households.

ENVIRONMENTAL HAZARD

Those elements in the physical environment which are harmful to people and caused by forces extraneous to them. It is a concept which is people-centred because the hazard does not exist unless we are affected. The hazard may be natural such as **flood** or quasi-natural such as smog where people and environmental processes combine, or people-made.

EPHEMERAL STREAM

A stream which only flows occasionally. Most streams in hot desert regions are ephemeral because they only flow after a storm which is a rare event. Then the **wadis** and **arroyos** become major streams for a short period as the **flash flood** passes down them. In temperate latitudes they are found in areas of permeable rocks such as chalk and depend on the seasonal height of the water table before they flow. This type of ephemeral stream is called a **bourne**.

EPICENTRE

◀ Earthquake ▶

EQUATORIAL CLIMATE

The climate associated with the **tropical rainforest biome**. It is dominated by equatorial air masses and the convectional rainfall associated with weak low pressure systems and thunderstorms brings at least 60mm per month of rainfall. The temperatures vary very little between the seasons remaining close to an average of 28–30°C, see Fig. E.2. Where there is seasonality in

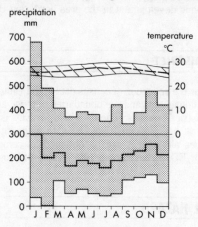

Fig. E.2

the rainfall pattern it will be due to distance from the equator and the passage north and south of the **intertropical convergence zone** and/or the effect of the monsoon.

EQUILIBRIUM LINE

An imaginary line on a **glacier** or an **ice sheet** which divides the **accumulation zone** from the **ablation zone**. See Fig. G.1.

ERG

◄ Sand sea ►

EROSION

Erosion is the second stage of **denudation** and is the physical process shaping and moulding landforms as a result of the work of running water, sliding ice, breaking waves and wind-borne material. **Chemical erosion** can occur both in the formation of coastal scenery and **karst** landscapes. Water, ice, waves and wind would not have any effect on a landscape without transported debris which actually erodes the surface. Preventing erosion depends on the transport of materials away from the point of erosion.

ERRATIC

1 A large piece of rock which has been transported by a glacier away from its area of origin and deposited in an area of different geology. Erratics provide us with clues as to the spread of glaciers during the **Quaternary period**.
2 A term used in statistics to denote values which vary considerably from the average.

ESKER

A **fluvio-glacial landform** produced by material deposited by **subglacial** or **englacial** streams. They form ridges of debris which wind across the country. Others believe that they are merely a delta form developed beyond the glacial front. They may be beaded, i.e. have wider sections produced by different rates of water input into the stream carrying the debris.

ESTUARY

The mouth of a river where the environment is produced by the unique combination of the fresh water of the river system with the salt water of the ocean system as the tides ebb and flow. It is the flow of water which scours the channels and which is more active in times of river flood when more water and more suspended particles are present to aid erosion.

EUROCENTRIC

Ideas, concepts and theories are described as eurocentric if they appear to be based on experience in the **developed world** and not the **developing world** to which they are being applied. **Rostow's stages of economic growth model** is a case where this criticism has been made.

EUROPEAN COAL AND STEEL COMMUNITY

This was set up in 1951 to help mining and steelmaking. Its role became more crucial as the coalfields and steel-making towns faced **deindustrialisation**. Today it administers grants to convert the local economy and to retrain the workers made redundant from the two industries. It is a form of international regional planning initiative.

EUROPEAN COMMUNITY

Formerly known as the European Economic Community (EEC), the modern association of West European states was founded in 1958 when the six member states, Belgium, France, Italy, Luxembourg, the Netherlands and West Germany became members following the Treaty of Rome in 1957. In 1973 Denmark, Eire and Great Britain joined, to be followed by Greece in 1980 and Portugal and Spain in 1986. In 1989 the Community was faced with the prospect that states in Eastern Europe might request membership following the marked shift away from socialist rule there. In addition Austria, Turkey, Sweden and Finland had also shown signs of wanting to join. The future size of the Community is therefore uncertain. The Community has developed a series of policies of which the **Common Agricultural Policy** is the most famous. Other policies exist for regions (CRP) 1973, transport and social needs (ESF) 1973. There is also a **European Investment Bank** (EIB) founded in 1958.

EUROPEAN INVESTMENT BANK (EIB)

The EIB was created by the European Community in 1958 to make loans to help economic development projects in less developed regions. It has provided much assistance to Italy, and also to the United Kingdom, which has received almost 20 per cent of its disbursements since 1973. More recently it has been involved in assisting development in Greece. It also provides funds for environmental improvements and to reconstruct areas after hazard events such as earthquakes.

EUROPEAN REGIONAL DEVELOPMENT FUND (ERDF)

One of the measures of the European Community to establish a mechanism for providing regional development aid throughout the Community. It has given its assistance mainly to Italy and Great Britain although more money now flows to assist the Mediterranean states through the Integrated Mediterranean Programme (IMP). ERDF funds are still sought for industrial and other economic investment in the **development areas** in Britain. This is because of the principle of **additionality**, i.e. an area must already be in receipt of aid from the national government to qualify for aid from the fund.

EUROPEAN TYPE

◄ Population-resource ratio ►

EUSTATIC CHANGE

A sea level change which has been caused by the alteration in the level of the sea. Most change has been caused by the abstraction of water to form the glacial sheets of the ice age or Pleistocene period. Some changes have been produced in more recent times by plate tectonics and the spread of the sea floor. Some people are of the opinion that the greenhouse effect will result in a melting of the ice caps which will return water to the oceans, making them rise over the coming decades.

EUTROPHICATION

The natural process of ageing in an aquatic ecosystem through the enrichment by nutrients. At the other extreme a water body may be described as oligotrophic because it has a low nutrient content and so productivity is low and it is unable to support much life. High altitude lakes are oligotrophic. The intermediate stage or mesotrophic is the state of many water bodies such as Austrian alpine lakes today or Barton Broad in 1800. Barton Broad in 1981 was verging on the state of eutrophication. There is a stage beyond the eutrophic called hypereutrophic, which is characteristic of a sewage outfall. Both the oligo- and mesotrophic systems can cleanse themselves because there is enough oxygen left in the water to decompose the waste. However, in those aquatic systems where there is a major input of industrial waste sewage effluent and agricultural fertilisers which are high in organic substances, cultural eutrophication takes place. Human action accelerates the process of eutrophication because the decomposition of the wastes needs increased amounts of oxygen. The amount of oxygen needed to decompose wastes in water is known as the biological oxygen demand (BOD). The higher nutrient levels cause plants and animals to grow more rapidly and, as they die, they starve the living plants of oxygen which is being used in the process of decomposition and decay.

EXFOLIATION

A form of mechanical weathering in which the outer layers of rock are split either by the expansion of salt crystals in the surface layers or by heating of the outer layers.

EXOGENOUS STREAM

A stream in an arid area whose supply of water comes from beyond the confines of the arid area. It is a permanent stream such as the river Nile, whose sources are in East Africa.

EXPANDED TOWN

◀ Town Development Act 1952 ▶

EXPANSIONIST POLICY

◀ Population policy ▶

EXPORT BASE ACTIVITY

◀ Base activity ▶

EXPORT BASE THEORY

An increase in the **basic activity** of a region will bring extra income to the region which can then be used either to increase the demand within the region for goods and services and/or to invest in further growth of the basic activity or other activities. Basic activity has a **multiplier effect** in that it leads to increased economic activity throughout the region. This is most easily measured by the **employment multiplier**. The removal of the economic base can lead to a decline in a region's economic activity. A good example of the effect of a rise and fall in economic base on the fortunes of a region is North East England.

EXTENSIVE AGRICULTURE

Farming which is characterised by low levels of labour and capital inputs per hectare and a low level of output. In contrast to **intensive agriculture** this system is characterised by large units of land holding, high levels of mechanisation and very little labour but with a high output per worker. The term is *relative* because cereal farmers in the UK are reagarded as *extensive* farmers whereas the cereal growers in N. America grow far less per hectare and make the British farms seem like *intensive* units. It is often practised in **marginal areas** where there is very little between the costs and revenues of a farm, and where the physical conditions are harsh. **Shifting cultivation** is regarded by some as being a form of extensive agriculture.

EXTERNALITIES

These can be *positive*, i.e. creating benefits, or *negative*, i.e. creating costs, and result from the actions of an individual or an institution acting in a way over which others have no control. Good gardens are a positive externality for house owners whereas a football ground may create negative externalities.

The strength of the externality normally declines with distance from its source. Therefore people will compete to be near the positive, e.g. a waterfront or a view. Conflict over externalities often results in **segregation**. Industry, offices and retailing also search for positive externalities, e.g. a Saville Row address for a tailoring firm or proximity to a Marks and Spencer store for a retail shop.

EXTRUSIVE VULCANICITY

The development of **volcanoes** and other landforms associated with the pushing out of volcanic magma onto the surface of the earth. The landforms which result may be the product of extinct activity, e.g. the Puy de Dôme, or dormant not having erupted in the recent past, e.g. Mount St Helens before 1980, or active. It may also result in **lava flows** such as the Deccan in India and **fissure eruptions** along fault lines.

FACTORS OF PRODUCTION

The essential prerequisites of any manufacturing process. They are LABOUR, CAPITAL, LAND and ENTERPRISE.

FALLOW

An agricultural practice which leaves land uncropped for a period to enable it to recover its fertility. It is a practice which was very popular in the past but it was costly because land was not being used. The soil was tended by ploughing and harrowing so that the decaying weeds provided nitrogen in the aerated soil where the processes of recovery could take place more rapidly.
◄ Von Thünens model ►

FAMINE

The severe shortage of food which may result from both physical and people-aggravated causes. Famines can be produced by:
  recurrent **droughts** as has happened in the Sahel;
  destruction of crops by **flooding** as in Bangladesh;
  destruction by **hurricane**, e.g. the West Indies at different times
  by the effects of pests such as the locust.
Overpopulation is also a cause because it leaves the agricultural areas more vulnerable to the effects of the physical causes noted above. Aid is one temporary solution to the problem of famine but aid does not produce the next year's crops unless part of that aid is fresh supplies of seeds and young animals. Many see famine as an increasingly frequent signal of the imminent advent of **Malthusian** conditions in the least developed countries.

FEDERAL STATE

A system of government which separates the government of a country from a second tier of regional or state governments. Power is thus divided between the two tiers. Many modern states are federations, e.g. USA, USSR, Canada and Australia. The Allies insisted that both West Germany and Austria became federal states after 1945. In the former case there are 11 second tier states or *lander* each with its own government, land capital and control over, e.g. education. Switzerland has been a federal country for much longer.

FERRALISATION

Intense transformation of soil material, typically in tropical environments. All salts and cations and also much of the residue silica are lost to the soil in solution. What's left in the soil is therefore: some quartz, 1:1 clay minerals, Fe and Al oxides. It is a very acid soil, with low base status and high levels of exchangeable aluminium.

FERTILISERS

Any inorganic or organic material of artifical or natural origin used to supply certain elements to the soil essential for plant growth. Farmyard manure is an example of an organic fertiliser, and N-P-K an example of an inorganic fertiliser.

FETCH

◄ Waves ►

FIFTH WORLD

Those parts of the world where the **Malthusian** checks of famine and pestilence are beginning to bite. Chad and Ethiopia are the countries most often cited in this category. They are the very poorest countries for many reasons, not least climatic failure and military expenditure which combine in the most disastrous fashion.

FILTERING

Filtering is the tendency for a house to be progressively occupied by lower socio-economic groups during its life, see Fig. F.1. Finally the property might be subdivided, demolished for redevelopment or **gentrified**.

central
business
district

I.

old housing
occupied by
low-status groups

better housing
occupied by
lower-middle
classes

better housing
occupied by
upper-middle
classes

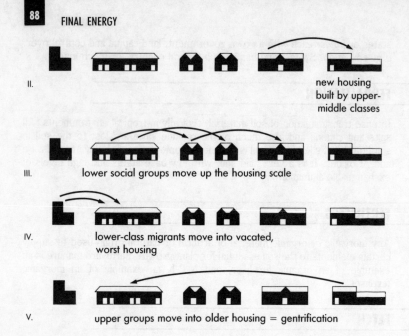

II. new housing built by upper-middle classes

III. lower social groups move up the housing scale

IV. lower-class migrants move into vacated, worst housing

V. upper groups move into older housing = gentrification

Fig. F.1 Filtering

FINAL ENERGY

The energy which reaches the consumer after distribution by electricity and gas grids, and oil pipelines and road tankers. There are always distribution losses and storage losses so final energy is less than secondary energy production.

FIRN

The accumulation of last year's snow on a glacier which is then buried by the next year's snowfall so allowing the glacier to increase its volume in the **zone of accumulation**. The firn is strictly that snow which remains after **ablation** during the summer months.

FIRST WORLD

◀ Developed world ▶

FISHING

A primary economic acitivity involving the exploitation of the **flow resources** of the oceans and seas. Fishing concentrates on either the sea bottom, such as demersal fishing grounds of the temperate continental shelves, or the pelagic, surface living species which tend to predominate in the tropics. Fish farming has increased in importance because a controlled environment can accelerate growth and is mainly used for the more expensive species such as trout, salmon and shellfish. The distribution of the major areas of fishing in the world can be related to the supply of nutrients either from the major rivers discharging into the seas or from the ocean currents and upwelling which carry nutrients from the nutrient rich waters of the polar regions.

FISSURE ERUPTION

The emission of lava from a fault or other line of weakness on the surface of the earth. Where this form of **extrusive vulcanicity** occurs it produces large lava plateaux such as the Antrim Plateau which is formed of the fluid, basic **lava** associated with such eruptions.

FJORD

◀ Glacial trough ▶

FLASH FLOOD

1 A very brief rise in the flow of a stream caused by a sudden input of water into the channel system. This will be the result of storms in desert areas which fill otherwise dry steambeds, flows through **wadis** and **arroyos** and deposits the transported materials in **alluvial fans**. Flash floods may also occur in temperate regions as a result of heavy rainfall so that **overland flow** replaces **infiltration**. Because the **discharge** is so great and the velocity so high, flash floods are capable of transporting a very large load in a short space of time. Stream flood is an alternate term.
2 A short duration **environmental hazard** following a rapid increase in precipitation, the collapse of some form of dam or even a people-made dam. In many cases the flood is made worse because of human actions and activities which, when combined with the vastly increased quantities of water, provide the ingredients for a hazard. The Rottingdean Flood of 1987 was one such example.

FLOOD

A form of **environmental hazard** caused by high magnitude of precipitation in a **catchment basin** which rapidly enters the stream system causing a rise in the river level and flooding once the bankfull capacity has been reached. The inundation of the land is not necessarily a hazard, but once it threatens life and livelihood it is a major and frequent hazard. Flood may also be caused by snowmelt, e.g. the Rhine in spring, and inundation by the sea as a result of a storm surge, or of abnormal weather conditions in a marine area adjacent to a low lying coast. In the latter case the floods in Bangladesh around the Ganges delta may be caused by typhoons in the Bay of Bengal and by flooding from the river itself. In such areas the ability of people to adjust to the hazard and the way in which they finally adjust varies considerably from the **developed world**. Human actions such as **deforestation** may lead to flooding.

FLOW LINE MAP

A map which has been designed to show the movement of commodities or people by making the lines show the flow proportional to the flow between the points on the map.

FLOW RESOURCES

◀ Renewable resources ▶

FLUVIO-GLACIAL LANDFORMS

◀ Outwash plain ▶

FLUVIOKARST

Landforms which have been produced by the combined action of fluvial, river processes and **karst** processes. Gorges with streams such as those in the Causses in southern France are regarded as fluviokarst features.

FOCUS

◀ Earthquake ▶

FOG

Fog is produced when condensation at or near ground level causes visibility to fall below 1 km. See Table F.1.

Type of fog	Season of occurrence	Areas affected	Mode of formation	Mode of dispersal
Radiation fog	Autumn and winter	Inland areas, especially river valleys and low-lying damp ground	Cooling due to radiation from the ground on clear anticyclonic nights in conditions of light winds	Heating of the ground by the sun or increased wind
Advection fog a) over land	Winter and spring	Often widespread inland	Warm air cooled by movement over cold land	Change in airflow or heating of the land
b) over sea	Spring and early summer	Sea and coastal areas	Warm air cooled by movement over cool sea surface	Change in airflow or heating of the coast
Frontal fog	All seasons	Inland, especially high ground	Warm air mass in contact with cold airmass in a weak circulation	Increase in intensity of the circulation or passage of front
Upslope or hill fog	All seasons	High ground	Low cloud forming below the summit of the hills	Change in circulation

Table F.1 Summary of fog types affecting the British Isles.   (*after Musk*)

FÖHN WINDS

Warm winds descending on the lee side of mountain areas. They are caused by rising air on the windward side cooling at the saturated adiabatic lapse rate and warming as it descends on the lee side at the dry adiabatic lapse rate which is a faster rate of temperature change. When such a wind blows, the rise in temperature can be up to 20°C in a few minutes and cause an avalanche threat. This type of wind is also known as the Chinook in the Rocky Mountains, Santa Anna in California and Zonda in Argentina.

FOOD AID

◀ Aid ▶

FOOTLOOSE INDUSTRY

An industry whose locational choice is relatively free because it depends on a wide variety of raw materials, e.g. motor vehicle assembly, or because its raw materials are readily available almost anywhere, e.g. brewing, or because the other factors of production such as labour are more important.

FOOTPATH EROSION

A form of ecological damage which results from the **ecological capacity** of the footpath and its immediate surrounds being exceeded for a period of time. It has to be stopped by diverting the footpath or the walkers, or by careful management of the course of the footpath combined with reseeding.

FOOT SLOPE

◄ Penck's slope replacement model ►

FORECASTING

The activity which is attempting to predict changes and their effects into the future. It can be applied to forecasting the economic changes and their effects on regions, the effects of a new shopping centre on older shopping centres and, more familiarly, to the prediction of weather patterns. Short term forecasting is generally more accurate for all types of activity and events than long-term forecasting. The process is an essential element of **planning** because it ennables the planner to extrapolate from the known to the predicted and to suggest the effects of a policy in terms of further demands on an area.

FOREST

An area of extensive woodland which can either be the natural vegetation or planted for its timber resources in particular. The process of exploiting the forest or forestry is a primary economic activity and involves the total management of the ecosystem which has been created. Forestry involves the total management of the trees from planting, **afforestation**, replanting or **reafforestation**, thinning, eliminating diseased species and felling for sale. Such management is imperative to preserve the **flow resource**. Siviculture is the cultivation of forest solely for its timber resource.

◄ Commercial tree crops ►

FOREST PLAN

A national plan to ensure that the forests of a country continue to provide revenue. This is essential for tropical countries anxious to guarantee revenue from the **flow resource**. It also involves managing the forests themselves. While only five African countries have a forest plan the forests will be threatened. Only 1 per cent of the African forests are managed despite these plans. Even if forests are managed, loggers can be bribed so that more is felled than is permitted by law. Forest plans normally involve:

i) replanting on a 1 for 1 basis,
ii) **reafforestation** of **marginal land** no longer needed for agriculture,
iii) improving the generic quality of the forest stock,
iv) using methods which opitimise the growth of the trees.

The type of plan will vary according to the location of the forest, e.g. coniferous forests in Sweden will be managed and planned differently from those in Malaysia although the basic aims are the same.

FORMAL REGION

An area whose uniformity is derived from any one characteristic such as geology and scenery in the Mendips, climate of the Mediterranean or land use in the Rhine gorge.

FOSSIL FUELS

The major energy resources of the **developed world** which include oil, natural gas and coal. It also includes peat, tar sands and oil shales but *not* **uranium** which is a radioactive metal. Ninety per cent of the global use of energy is based on fossil fuels.

FOURTH WORLD

Those least developed countries whose economies are so poorly developed that they are caught in a vicious cycle of increasing poverty. Many of the twenty poorest economies according to the World Bank such as Nepal, Bangladesh and Zaire fall into this category.

FRANCHISING

A trend in retailing where individuals raise the capital to establish an outlet and the shop design is controlled by the franchiser. The franchisee pays a fee for every item sold to the franchiser. McDonalds and other fast food outlets are the most common franchisers, although the system is now entering other forms of retailing such as clothing, jewellery and even travel agencies.

FRANK'S DEPENDENCY THEORY

The theory examines the **development of underdevelopment** in terms of the domination of the capitalist world over global economic development and the **dependence** of the developing world. Surplus value of production (i.e. profits) are taken by developed countries and particularly by the transnational and multinational companies operating from those countries and this is a normal process of capitalism. To gain these profits the dominant capitalist world has used the labour and resources of the **developing world**. Aid and loans merely increase dependency and cause countries to have massive debt problems such as those faced by Brazil and Mexico in the 1980s. Frank maintains that the only way to break this dependence is a new social and economic order. The theory was developed in 1969.

◀ Colonialism ▶

FREE-FACE

◀ Slope form ▶

FREE PORT

An area at a port or an airport where goods can be landed, processed and re-exported without having to pay **tariffs**. This enhances the trading position of a port and also enables traders to store goods awaiting the best time to import them into a country. Hamburg has such a facility and the British government has established a series of freeport zones, e.g. in part of Southampton Docks, but they have had limited success compared with similar schemes in Europe and the Far East.

FREE TRADE

Trade which can take place freely because of the absence of **tariffs**. There are also no quotas on trade. Some countries develop **free ports** in order to encourage the development of industries free from tariffs which are levied on normal imports and exports.

FREEZE-THAW

A form of **mechanical weathering** in which water freezes in fissures, splitting rocks by increased pressure. It thaws and refreezes so progressively weakening the rock.

FREQUENCY DISTRIBUTION

The scatter of observations of recorded data on a graph in the form of **histograms** or grouped frequencies. The description of the graphical form of the grouped frequencies relates the form of the frequency distribution to that of the normal symmetrical distribution curve and can be bimodal, leptocurtic, mesokurtic, platykurtic and skewed.

FRIEDMANN'S CENTRE–PERIPHERY MODEL

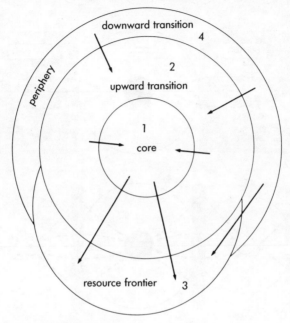

Fig. F.2 Friedmann's centre–periphery model

This model shown in Fig. F.2 argues that development must spread from the core regions or countries to the less developed periphery. At the global scale there are four divisions;

core regions, W. Europe, N. America and Japan,
upward transition regions, SE Brazil, S. Korea, Israel and Kuwait,
resource frontier regions in both the developing and developed worlds,
e.g. Siberia, Alaska and NW Brazil,
downward transition regions, e.g. Chad and Ethiopia.

Friedmann also developed a model, see Fig. F.3, which is based on the core-periphery model to suggest the stages through which a national economy would pass on its route to development and how these stages would affect the spatial structure of that economy. Stage 1 is a pattern of isolated separate cities. Stage 2 is a phase of incipient industrialisation when the core dominates the economy and draws its wealth/raw materials from the periphery. The core has become a **primate city** in a colony. Stage 3 sees the development of second order centres in the periphery with a series of peripheries existing between the metropolitan centres. All of these centres assist in the growth of the national economy. Stage 4 is an interdependent system of cities in a fully integrated economy. The peripheries have disappeared and growth is

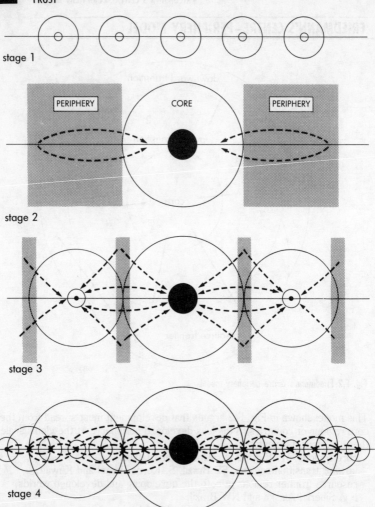

stage 1

PERIPHERY   CORE   PERIPHERY

stage 2

stage 3

stage 4

Fig. F.3 Friedmann's development model

maximised throughout the country. This model was applied successfully to development in Venezuela.

FROST

Frost occurs when the temperature measured in a Stephenson screen is below freezing point. If dew is frozen it is called white frost. When it occurs on roads it is a traffic hazard known as black ice. Frost is a hazard to market gardening if it comes late in winter just as early crops are at their most vulnerable to temperature changes. Frost has destroyed coffee crops in Brazil and even the oranges in Florida in extreme years.

FROST CRACKING

A crack opened up in the soil by ice which will grow into an **ice wedge**.

FROST HEAVE

A **periglacial** process in which the soil rises into a dome as a result of the pressures of the freezing water beneath the surface. It is a form of **congeliturbation**.

FROST SORTING

A general term referring to the **periglacial** processes which result in the formation of **patterned ground** including **frost heave, thrust** and **freeze thaw**.

FROST THRUSTING

A **periglacial** process in which the soil moves laterally as a result of the pressures of freezing groundwater. It is a form of **congeliturbation**.

FUMAROLE

A small vent on the earth's surface which emits steam. They are associated with zones of declining volcanic activity.

FUNCTIONAL REGION

Regions of interdependent parts such as a **drainage basin** or a city region. They make the ideal regions for planning and have been widely used for this purpose.

FUTURE SUPER-ADVANCED SOCIETY

◀ Mobility transition model ▶

GARDEN CITY MOVEMENT

The movement founded by Ebeneezer Howard to promulgate the development of the garden city. It developed Letchworth and Welwyn, influenced the development of Hampstead Garden Suburb and had a profound influence on suburban developments in Europe where *cités jardins* and *gartenstadt* were built. It is the indirect forerunner of the **new town**.

GARDEN FESTIVALS

Five garden festivals have been held or planned, e.g. Liverpool, Glasgow, Stoke and, in 1992, South Wales. They are attempts to provide, in derelict areas of cities, a new townscape which might remain after the festival. The experience of Liverpool after the festival suggests that their impact may only be short term and they are merely a one year palliative to urban decay.

GATEKEEPER

A term used in **urban geography** to identify those managers who may actually control the allocation and distribution of benefits to the population or to households. Studies have suggested that the allocation of mortgages depends partially on the assessment of the applicants by managers and by credit agencies who by their very rules of operation are managing the allocation of housing to particular people. The allocation of council housing has seen other rules which have the effect of segregating certain groups of people to the less desirable areas. In this way the urban managers are controlling the allocation of space.

GATT

This is an acronym for the General Agreement on Tariffs and Trade, an international agreement to assist in the liberalisation of trade between countries which has attempted to manage the progressive reduction of **tariffs** on a variety of products. It is a means of moving towards **free trade**.

GENERAL FERTILITY RATE

The number of births per year per thousand women aged 15–45 years old. These may be measured more accurately as the **age specific fertility rate**, the number of births per year per thousand women 25–29, 30–34 etc. Fertility varies with age of the mother, the number of existing children in the family unit, the occupation of the parents and especially the mother, medical knowledge, fashions in family size, the type of area and the social pressures to have children of a particular sex.

GENTRIFICATION

The movement of higher socio-economic groups into an area following its renovation, often as a result of it being a **conservation area**. The process normally involves the outmigration of existing residents either bought out or forced out by landlords (though the latter is not normal today). As a result some of the run down districts of cities are revitalised, e.g. Wapping and Islington in London. It has often come at the end of a process of downward **filtering** of the property.

GEOMETRIC MEAN

A measure of central tendency which is little used. At A-level, the **arithmetic mean** is preferred.

GEOMORPHIC HAZARD

A type of **environmental hazard** or natural hazard which is the product of the processes of erosion and weather such as **avalanche, earthquake, flood,** landslide, **tsunami, volcanic** eruption **shifting sand** and **erosion**.

GEOMORPHOLOGY

The study of the form of the land surface. It includes the study of the processes which have produced landforms and the history of those landforms. The processes are those of gravity, water as both glacial and fluvial forms, waves and wind. The processes are affected by other environmental variables, namely; geology, climate, vegetation and base level.

GEOTHERMAL ENERGY

◀ Energy ▶

GEOTHERMAL POWER

◀ Energy resources ▶

GERRYMANDER

An American term used to signify the manipulation of electoral boundaries in order to ensure political support for a particular group. It is used today even where there has not been *deliberate* action, but where the boundaries seem to favour a particular party.

GEYSER

A discharge of superheated water and steam from an underground chamber which is heated by volcanic rocks in the crust and the pressure of the head of water. The geyser and thermal springs often leave deposits around the vent. Geysers are a tourist attraction and also have led to the areas where they occur being used for the development of **thermal power**, e.g. Lardarello, Italy and Rotorua, New Zealand.

GHETTO

Originally the name given to the area where the Jews lived in the mediaeval city (named after a district of Venice). More recently it is the term given to extreme **segregation** of a group who are socially or economically deprived. It normally results from fear, prejudice and even economic sanctions in the host society. Ghettos form because of *avoidance* which enables the community to develop its customs best in an area where others are absent, *preservation* of the culture especially in the home area and the *defensive* needs of the community especially to protect the newcomer. Ghettos grow by spillover, the gradual spread of an area dominated by a minority group, by leapfrogging, which is the more sudden attempt to colonise a new area and in response to local pressures to move, e.g. renewal policies resulting in **gentrification**. It is worth remembering that ghetto does not mean a slum; some ghettos are known as 'gilded ghettos' because they are the residential areas of the more affluent members of a minority group. The most extreme cases of ghettos today are enforced by law in South Africa and include areas such as Soweto.

GLACIAL TROUGH

The overdeepended valley which results from glacial erosion in an upland area. It is normally bounded by **hanging valleys** and **truncated spurs**. It may contain **rock steps** and **ribbon lakes**. If the trough has been flooded by the sea as a result of **isostatic** and/or **eustatic changes** in sea level then a **fjord** is the name given to the submerged trough.

GLACIER

An area of ice as shown in Fig. G.1 which is of limited width and normally

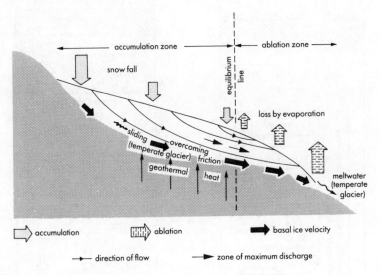

Fig. G.1 The glacial system

flowing in a valley or confined to an area by major landforms such as mountain ridges. It can be seen as a **system** with the inputs of ice in the **accumulation zone** and the outputs of water and material in the **ablation zone**.

GLASSHOUSE CROPS

Market gardening takes place in glasshouses where the plants may be protected against adverse weather conditions and can be given artificial light and heat to stimulate early or out of season growth. The glass also traps radiation and therefore higher temperatures which encourage growth. The glasshouse industry has been able to refine its techniques so much that the new **biotechnologies** and **hydroponics** have benefited the growers in recent years. The major crops are tomatoes, lettuce and cucumbers although modern exotic crops such as peppers, courgettes and other semi-tropical fruit and vegetables are now being grown. Flowers and pot plants are also major products.

GLEY

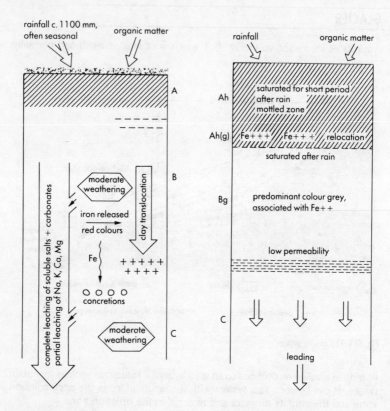

Fig. G.2 Groundwater gley

Fig. G.3 Surface-water gley

A soil type dominated by the process of gleying, see Figs. G.2 and G.3. Poor drainage results in the chemical reduction of iron and other elements to give grey colours typically seen as mottles.

GLEY PODSOL

A podsol soil with a B_s horizon, see Fig. G.4. However, the water table may periodically be found in the B_s so producing gley features. Hydromorphic podsol under the Food and Agricultural Organisation (FAO).

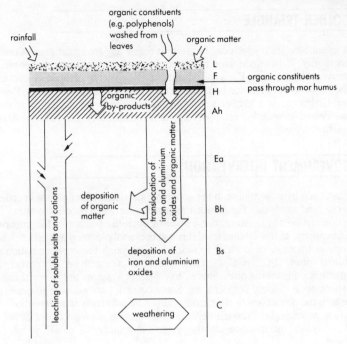

Fig. G.4 Podsolisation

GLOBAL EQUILIBRIUM

◄ Limits to Growth Report ►

GLOBAL SHIFT

A term devised by Dicken to indicate the developing patterns of industrial location where **transnational corporations** control the international distribution of manufacturing and services. The resulting patterns of economic activity have shown marked changes in the location of production and services which have resulted from the **international division of labour** and the **japanisation** of economies.

GNAMMA HOLE

Pits on the sides of rock domes in the desert which have been hollowed out by weathering.

GOLDEN TRIANGLE

A journalistic term which can be used to define any triangular area of relative prosperity. It has been used to define the core of the European Community between Amsterdam, Paris and Milan, and also the core of the development in the 1980s west of London bounded by lines joining Heathrow Airport, Reading and Guildford. At a smaller scale it has been used to retail analysts to define the desirable area for retail development bounded by the M1 and M10 in Hertfordshire.

GOVERNMENT INTERVENTION

The action by government in the economic and social life of a state in order to rectify and ameliorate the greatest inequalities between people and areas. The action can be at the international level through the policies of the **European Community**, at the national level through **regional planning** and other policies such as taxation, or even at the local level through the implementation of a **district plan**. In capitalist states there is a gradation of government intervention from the more liberal, free market systems to the social-market economies and social democracies. Some countries are actively attempting to reduce the distortions to the market caused by government intervention. The UK is an example, although here the intervention is probably not direct but the indirect manipulation of the economic climate to assist people and areas.

GRABEN

A downfaulted area which gives rise to a valley such as the Rhine Rift valley between Basle and Bingen. **Erosion** will eventually remove the rift.

GRANULAR DISINTEGRATION

A form of **mechanical weathering** caused by either the freezing of pore water or by the expansion and contraction of the rock due to solar insolation, resulting in the break up of the rock.

GRAVITY

The force which is exerted by the earth and its rotation in proportion to the mass of the earth and an object, and inversely proportional to the square of the distance between the earth and the object. It is the force which moves glaciers on a slope, acts to assist the downslope movement of rocks and regolith in earth movements and overcomes the forces of **transport** when **deposition** occurs. The gravity model is also used within human geography to determine the **sphere of influence** of settlements and in Reilly's law of retail gravitation to determine the theoretical patterns of shopping behaviour.

GREEN BELT

This is an area of rural or semi-rural land surrounding a city on which further urban development is to be prohibited or severely restricted. It was part of Howard's **garden city** idea and became law in the 1938 Green Belt Act. The original green belt was around London to check its growth and spread, but others have been created to keep two towns apart, e.g. Bath and Bristol or the Ruhr cities. In these cases it is not so much a belt as a buffer. Green wedges, corridors, hearts and zones also exist as variants on the same policy, see Fig. G.5. They can help preserve the special character of a town and provide for recreational needs. Green belts are seen as restrictive by property developers who would like to use land for housing and out-of-town retailing in particular because they resent the way that hospitals and mineral workings are permitted uses. Transport routes such as the M25 have used green belt land. Restricting development raises house prices and forces out the local population in favour of the affluent commuter who then fights to maintain the green belt.

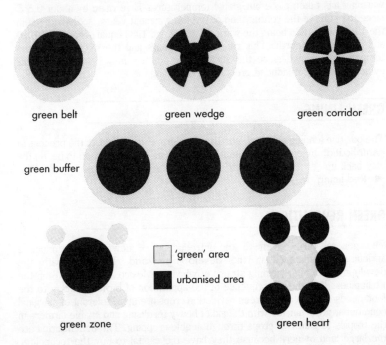

Fig. G.5 A classification of 'green' planning

GREENHOUSE EFFECT

Atmospheric gases are almost transparent to solar radiation wavelengths but the same is not true for terrestrial radiation which is partially absorbed and then re-emitted, some of that emission being back down towards the surface. The most important greenhouse gas is water vapour, but carbon dioxide, methane and nitrous oxide also act as greenhouse gases. The concern about the greenhouse effect is not that it exists, as the world would be much colder without it, but that human acitivity is adding to it and may cause a significant change in world climate.

Carbon dioxide is the gas most often mentioned in connection with the greenhouse effect and human activity has increased the concentration of CO_2 from perhaps 280 parts per million (ppm) in the 18th century to about 360ppm today with the bulk of that increase coming in the last 30 years. Carbon dioxide is formed when burning **fossil fuels** and when clearing forests by burning. Methane and nitrous oxide concentrations have also been increased by agricultural activity. Other potentially important greenhouse gases are the chlorofluorcarbons (CFCs) used in freezers, refrigerators and aerosol sprays. The CFCs are much more effective greenhouse gases than carbon dioxide: their concentration in the atmosphere is small but increasing.

As yet it is difficult to demonstrate that human-generated greenhouse effect warming has taken place but global temperatures have risen by about 0.5°C since the start of the century and of the ten warmest years, six have been in the 1980s with 1988 being the warmest, 1987 and 1983 equal second and 1989 and 1944 equal fourth. This run of warm years and the magnitude of the temperature change since the start of the century may be the first clear evidence of human-induced greenhouse warming.

GREEN LINING

The positive lending policy of mortgage lenders which supports the process of **gentrification** by ensuring plentiful loans to areas which are seen to be 'on the way back up'.
◀ Red lining ▶

GREEN REVOLUTION

The general term given to the introduction of new, more productive agricultural techniques in the **developing world**. It began with the development of high-yielding strains of wheat in Mexico and rice (IR8) in the Philippines. It is the consequence of the application of **biotechnology** to the food needs of the world. Green revolution crops are more tolerant of marginal conditions, grow quickly with the aid of heavy fertilising and enable farmers in the tropics to gain two crops through double-cropping. The beneficiaries are the large land owners because they have the capital to buy the technology whereas the poorer peasant is disadvantaged.

GROSS DOMESTIC PRODUCT (GDP)

GDP is the total value, excluding that of foreign workers' remittance, which has been produced by a country's economy. It is not as useful as GNP for measuring a country's income or wealth although it can be used as an **development indicator**. The percentage of the GDP from various sectors of the economy does indicate the relative strength of the sectors and is a further indicator of development.

GROSS NATIONAL PRODUCT (GNP)

GNP is the total value of all that has been produced by industry and services in a country by its resident population or transferred to the country by its own nationals from abroad. Foreign worker earnings are excluded. When expressed per capita it is a summary **development indicator** and measure of economic prosperity which is usually expressed in US dollars. The global pattern is shown in Fig. G.6.

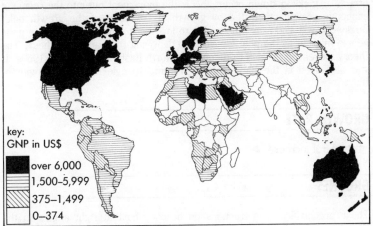

key:
GNP in US$

over 6,000
1,500–5,999
375–1,499
0–374

Fig. G.6 World pattern of GNP per capita

GROUNDWATER FLOW

The movement of water which has percolated through the soil regolith to the rocks beneath; it may contain water which has risen from below. It is generally a downslope movement although the lithology of the rocks will influence its movement. The source of much water supply from wells, it is essential for most weathering processes because it occupies the spaces and pores in the rocks and contains the agents of chemical weathering derived from the passage of precipitation through the soil.

GROUNDWATER GLEY

A soil showing features associated with the process of gleying in the sub-surface horizons. This results from the lower part of the profile being within the fluctuating water table zone. See Figs. G.2 and G.3 on page 102.

GROWTH POLE THEORY

Developed by Perroux as an economic theory, growth pole theory was given a spatial context by Boudeville. Its essence is that economic growth does not appear everywhere at once, but at points or **development poles**. Growth then spreads from these poles to the whole economy. **Leading industries** are the large companies which dominate the growth pole. Very often these are grafted onto an existing settlement such as Bari in Italy. The initial growth leads to **polarisation**, the rapid growth of other economic activities in the growth pole consequent upon growth in the leading industry. **Agglomeration** economies are the consequence for the growth pole. Eventually, **spread effects** or **trickling down** occur; this is the movement out into the region of economic growth stimulated by growth at the heart of the growth pole. **Backwash** will also cause the region to lose out to the growth pole, e.g. as people migrate to the pole for work so depriving rural areas of labour. The theory influenced policies to regenerate North East England in the 1960s. It was also used in the Bari–Brindisi–Taranto triangle of the Mezzogiorno, Italy.

GROWTH RATE

◀ Natural increase ▶

GROYNES

These are artificial constructions on beaches built at right angles to the shoreline, to arrest the movement of material along the beach by **longshore drift** and to maintain the material on the beach. They also assist in the prevention of cliff erosion by maintaining a store of material which dissipates the force of the waves before they reach the cliff foot.

GRYKES (or GRIKES)

The fissures which develop along joints in **carboniferous limestone**. The joints are widened by **chemical weathering** and form a lattice of ruts up to 10–20cm wide and 50–70cm deep which widen further where the grykes intersect. The upstanding blocks are called **clints** and together they form a **limestone pavement**.

GULLY

A severe form of erosion resulting in the formation of a channel. There is a concentrated but intermittent flow of water in the gully usually during, and immediately following, heavy rain. A gully is deep enough to interfere with, and not be removed by, normal ploughing and agricultural activities.

HAMADA

A coarse stony desert created from the splitting of rocks by **insolation weathering**. The term hamada also refers to the bare rock areas of deserts.

HAMILTON'S SOCIALIST CITY MODEL

Hamilton attempted to look at the structure of the socialist cities in East Europe and constructed a model which distinguished between the pre-socialist urban area and those areas built since 1945, see Fig. H.1. Emphasis is placed

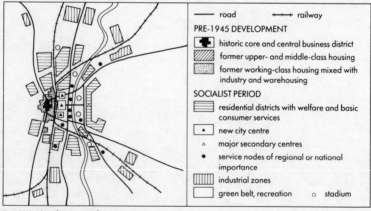

Fig. H.1 Hamilton's socialist city

on those elements of socialist culture which give the city its distinctiveness such as the new city centre with its area for military parades and the stadium. Perhaps this model will become obsolete after the events of 1989/90.

HAMLET

A small cluster of rural dwellings normally without any service or social functions such as a church. A hamlet is more than an isolated dispersed house, but smaller than a **village** and the term is a rather imprecise one as a result.

HANGING VALLEY

A tributary valley to the main glacial valley which joins at a higher level than the main valley floor. This is caused by the greater erosion and overdeepening of the main valley by its glacier so that the tributary valley is left at the higher level once the glacier has retreated. Streams in hanging valleys have a potential for hydro-electricity.

HARDPAN

A hard layer within the B horizon of the soil which has been cemented together as a result of **illuviation** and **leaching**. In Great Britain it is associated with **podsolic** soils and is called ironpan because the sand and gravel has been bonded by ferric salts. It may also be claypan from washed water down eluviated clay particles and moorpan from washed down and redeposited humus.

HARDWARE

The equipment used in the collection and processing of data. It can refer to the data logger, the computer, the printer, the digitising pad as well as the cameras used in **remote sensing** and aerial photography.

HARRIS AND ULLMAN

Key

1 central business district
2 wholesale light manufacturing
3 low-class residential
4 medium-class residential
5 high-class residential
6 heavy manufacturing
7 outlying business district
8 residential suburb
9 industrial suburb

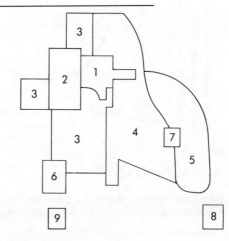

Fig. H.2 Harris and Ullman's multiple nuclei model

The multiple nuclei model, see Fig. H.2, was developed in 1945 on the basis of an initial study of Calgary, Canada. It recognises the complexities of city development and the fact that many settlements grow up around pre-existing village nuclei which become swallowed up in the spread of suburbia. As a result, an urban area will be structured according to the nature of these nuclei so that an upper class district might focus upon the most attractive village nucleus. Other nuclei grow up because certain functions repel one another or because the price of land forces certain uses to find less costly locations.

HAWAIIAN ERUPTION

◀ Volcano ▶

HAZARD ADJUSTMENT

People react to a hazard event which causes disruption so that there is an observable life cycle of a hazard with a recovery time as an area and its people return to normal, see Fig. H.3. Public authorities also adjust to a hazard in that

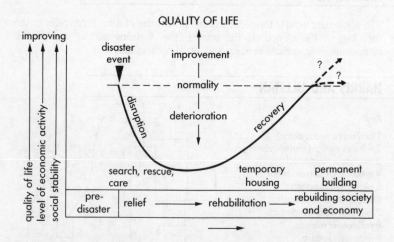

Fig. H.3 Adjusting to a hazard event

they must evaluate the threat from the hazard and then choose a path of adjustment which the society in the area can afford, see Fig. H.4.

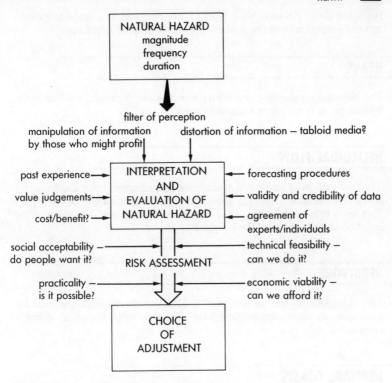

Fig. H.4 Pattern of official adjustment to hazards

HEAD DEPOSITS

Large areas of **solifluction** material which was moved during periglacial conditions. It can be found in the bottom of **dry valleys** in the South Downs where it is called **coombe** rock. It is also found on raised beaches in, e.g., Devon.

HEATH

Areas of relatively low land in the latitudinal belt 50–60° N & S which are dominated by ling, gorse and poor grasses. The sere evolved as a result of the burning of woodlands on areas of poor sandy, **podsolic** soils during prehistoric times. The plants are adapted to the **leached** sandy soils with their thin surface layer of **peat**. Much heathland has been developed for housing and recreation such as golf courses although large areas do exist especially on the reworked glacial outwash of the North European Plain, e.g. Luneburg Heath and the Veluwe area of the Netherlands. It can also be called *geest*.

HEAVE

The alternating expansion and contraction of loose rock pieces which move downslope. It is a form of **mass movement**.

HELICOIDAL FLOW

The flow pattern of water in a meandering stream which takes the form of a corkscrew-like flow within the general downstream movement of the water. This leads to erosion on the outer bank of a **meander** and deposition at the **point bar** on the inside of the meander.

HERBIVORE

The organisms at the second **trophic level** which consume the **primary producers** to obtain energy and food. They are also known as primary consumers.

HERITAGE COASTS

Areas of outstanding coastal scenery in the UK which were established by the Countryside Commission. They are designated as a part of the **struc-ture plan**. The first were established in 1970 and include the Dorset coast and much of the coast of Cornwall.

HIERARCHY

A series of ranked tiers of places, based on the interdependence between ranks although many see the relationship as dependence on the tier above. Normally the number in each rank diminishes with each move up the hierarchy. It is a concept which is used in **Christaller's central place theory** and in studies of settlements and their retail patterns.

HIGH STATIONARY STAGE

◄ Demographic transition model ►

HINTERLAND

A term derived from the German which refers to the area behind a place. It is used to refer to the land area from which a port derives its trade and to denote the urban field or **sphere of influence** of a city or town.

HISTOGRAM

A **bar chart** where there are two quantitative scales with the Y axis showing frequency and the X axis the size classes or values, see Fig. H.5.

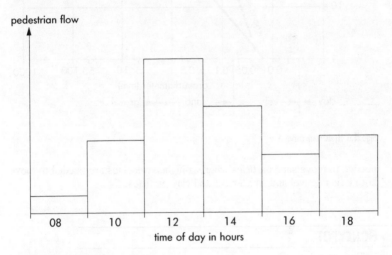

Fig. H.5 Pedestrian flow

HISTORICAL GEOGRPAHY

The geography of times past and the way in which past activities have helped to shape the landscapes of the present. Most of the landscapes and townscapes that are the subject of geographical investigation have been formed in the past; a decision to locate a factory might have been made ten or one hundred years ago. The morphology of a town is also the product of its evolution through time, a factor recognised by Burgess amongst others.

HJULSTROM CURVE

A graph based on empirical research which shows the critical **erosion** velocity needed to move particles of different sizes, see Fig. H.6. Lower velocities are

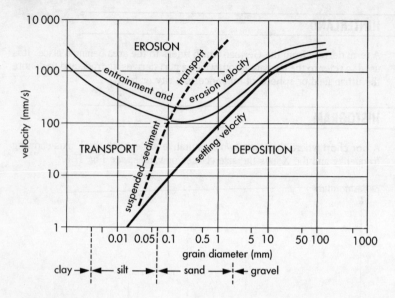

Fig. H.6 Hjulstrom curve

needed to move sand particles whereas higher velocities are needed to move both larger gravel and smaller silt and clay particles.

HONEYPOT

A particular location which has the effect of drawing people to it. The term is used in **recreation** geography for major tourist nodes which cause a cluster of visitors in an area with all the consequent problems of crowding, negative amenity and environmental damage. Honeypots may be rural, e.g. Hay Tor on Dartmoor, or the National Motor Museum in the New Forest, or urban such as the Eiffel Tower, or the Tower of London.

HOOVER'S THEORY OF INDUSTRIAL LOCATION

A theory of location devised in 1948 which is based on the assumption that transport costs taper or rise in steps with distance rather than in proportion to distance. He also noted that where different transport systems meet, goods must be transhipped and that additional costs are incurred. These extra costs may be avoided if a plant is located at the transhipment point which is a break of bulk point where goods are transferred from sea to rail or road.

HORIZONTAL LINKAGE

A concept in industrial geography which refers to the interdependance between firms who all supply one plant, often with different products. It is also called convergent linkage.

HORTICULTURE

The growing of fruit, vegetables and flowers, often in nurseries for commercial gain, which is a form of **intensive agriculture**. Today the terms **market gardening** and **glasshouse crops** are used to indicate the same type of cultivation.

HOT DESERT CLIMATE

The climate associated with the areas beneath the main tropical continental air masses with their subsiding air of the Hadley cell. They are associated with high temperatures although the daily variations in temperature can be extreme. Precipitation is negligible and comes from occasional storms in summer although the pattern of these is almost random and undependable. The **desert biome** is a product of this climate.

HOTELLING'S MODEL 1928

A model which examines demand in a linear market such as a beach, see Fig. H.7. Stage 1 shows the socially optimal location where the market is shared, stage 2 the relocation of X to increase its market share and stage 3 the competitive equilibrium which evolves as a result of Y's strategy to combat X's extension of the market area.

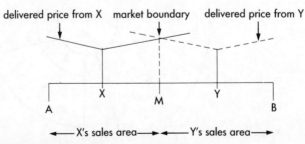

a) socially optimal locations

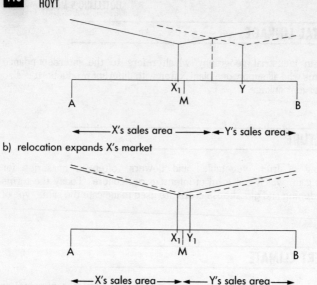

b) relocation expands X's market

c) competitive equilibrium

Fig. H.7 Hotelling's model

HOYT

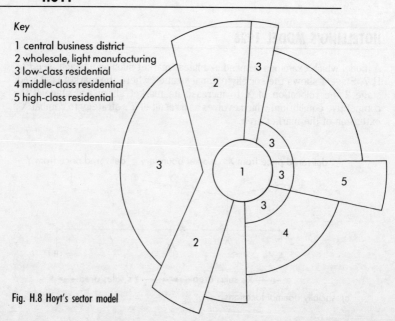

Key

1 central business district
2 wholesale, light manufacturing
3 low-class residential
4 middle-class residential
5 high-class residential

Fig. H.8 Hoyt's sector model

Hoyt's sector model of a city was developed in 1939. In contrast to **Burgess** he believed that the structure of a city was conditioned by the pattern of communications which radiate out from the centre, see Fig. H.8. The access along these routes varied and so different uses are attracted to each sector. The evidence which he used was rental and housing data. He showed that the high-class residential area was located in the most desirable part of the city while the lower socio-economic groups are housed adjacent to the industrial areas for ease of access. Industry is clustered along the rail routes. The model was initially based on Chicago although it was confirmed by studies of other American cities. It is also a product of its time, and subsequent studies such as that by **Mann** have refined the model.

HUMAN GEOGRAPHY

The study of the spatial patterns which arise as the result of the actions of people as decision makers. It can be subdivided into separate fields of study either on the basis of the subject matter, e.g. **economic geography** or on the basis of the approach or paradigm, e.g. **behavioural geography**. Other subject based subdivisions include;

cultural geography
historical geography
industrial geography
population geography
political geography
transport geography
settlement geography
rural geography
social geography
regional geography

Other approaches include **spatial analysis, structuralism, Marxist, humanistic,** and **welfare**. It can be studied at all scales from the local to the global although there are no strong links in terms of concepts which can be applied convincingly at all scales.

HUMANISTIC GEOGRAPHY

An approach to the study of geography which places human awareness and creativity at the centre of our studies. It was a reaction against the objectivity of the **quantitative revolution** and intended to make geography be about people and be concerned for the well-being of people. It places great stress on the nature of place and the way in which both place and space stimulate people.

HUMAN RESOURCES

The population of an area and its skills and talents.

HUMMOCKY GROUND

A landform characteristic of an area's **periglacial** activity where the ground has been pushed upwards by ice beneath the surface and by the movement of stones in the soil. Hummocks are generally no more than 3 metres in diameter and about 75cm high.

HUMUS

A stable state of soil in which organic matter remains as the end byproduct from the decomposition of added vegetation and animal residues.

HUNTING AND GATHERING

A primitive form of human lifestyle which depends on gathering food supplies from an ecosystem and relying on the ecosystem to replace the supplies naturally. Even such simple practices can result in resource depletion and the alteration of the ecosystem if population pressure results in overhunting. Bison were overhunted on the Canadian Prairies almost to the point of extinction.

Hunting may also be for profit, and more and more hunting activities are being banned world wide to help maintain species. Seals are no longer hunted for their pelt, elephant hunting is being slowly eradicated and all countries except Japan have banned most forms of whaling. In all these cases attention has been drawn to the contribution to the particular ecosystem that each of these mammals makes and the consequences for the system of its extinction.

HURRICANE

Hurricanes occur in late summer, between 5° and 15° from the equator, over western areas of oceans with sea temperatures of 27°C or more and with high pressure in the upper air. They are a violent means of energy transfer in the atmospheric system, see Fig. H.9.

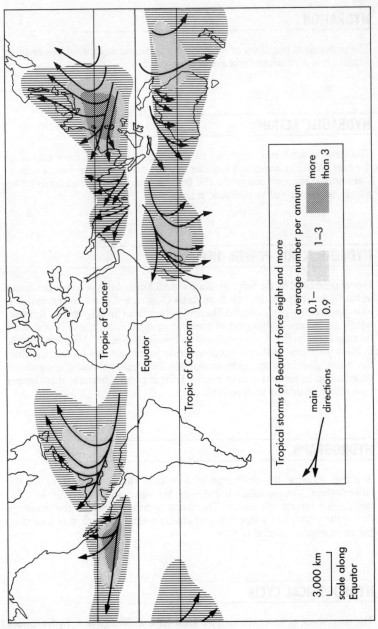

Fig. H.9

HYDRATION

The physical binding of water molecules to particles, molecules, ions or other matter. It is a restricted process of weathering.

HYDRAULIC ACTION

The force which water exerts on rocks which excludes the force applied by the load. Water in waves will force air into the cracks and crevices of cliffs so causing the disintegration of the cliff face. Streams flowing past a river bank also remove material by hydraulic action.

HYDRO-ELECTRIC POWER (HEP)

Power generated by the force of falling water. Most of this power is generated in North America and Europe from dams constructed in the mountains. It is also possible to obtain hydro-electric power from multi-purpose schemes which are primarily designed to improve navigation on major rivers. Such barrages exist in the Rhine gorge and on the rivers Neckar, Main and Moselle in West Germany. In the **developing world** major barrages on rivers such as the Niger (Kanji), Volta (Akosombo) and Nile (Aswan) make a significant contribution to national energy needs. HEP is popular because it is cheaper than either oil or coal fired power.

HYDROGRAPH

A graph showing the **discharge** of a stream channel at a place on the streamcourse. The graph will measure the discharge over time in cubic metres per second or cumecs. The storm hydrograph shows the discharge during the passage of a storm and its effects on the streamflow at a point on the stream over a period of time.

HYDROLOGICAL CYCLE

The movement of water through the **atmosphere, lithosphere, hydrosphere** and **biosphere**. It can be expressed both as a system and as a diagram, see Fig. H.10, which illustrates the cycle.

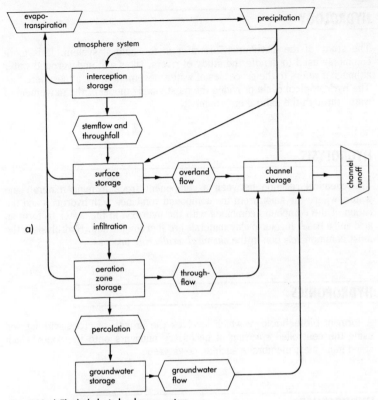

Fig. H.10 a) The hydrological cycle as a system

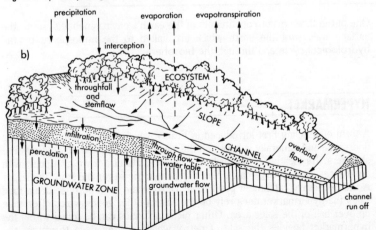

Fig. H.10 b) The hydrological cycle for groundwater flow

HYDROLOGY

The study of the distribution of water in the natural system. It is more commonly used to denote the study of rivers, lakes and underground water although it refers to the processes of water movement in the atmosphere too. The **hydrological cycle** provides the most simple model of the movement of water through the natural environment.

HYDROLYSIS

The chemical reaction between a component (rock forming mineral) and water, where the anion from the compound combines with hydrogen and the cation of the compound combines with the hydroxyl in the water to form an acid and a base. Insoluble clay minerals are thereby formed. Hydrolysis is the most common reaction in the *chemical weathering* process.

HYDROPONICS

A form of **biotechnology** which involves the growth of crops without soil using the controlled spraying of necessary nutrients onto the roots which stand in an inert membrane such as polystyrene.

HYDROSPHERE

One of the three major constituents of the earth's environmental system: the surface waters of the earth which are linked to the atmosphere by the **hydrological cycle** and through the **biosphere**.

HYPERMARKET

A form of retailing first introduced in France which is defined in terms of its floorspace and generally is based on the sale of mainly foodstuffs to car-borne shoppers. Its location is normally on the outskirts of a town at a point with good access from a large catchment population; the larger the store the larger is the necessary catchment population. Despite the main purpose being food sales the hypermarket does rely on the sale of other goods which often take up over half of the sales area. Other **niche** activities can be found within the hypermarket besides the sale of petrol which is often used to attract the shopper with cheap offers.

HYPOTHESIS

An unproven proposition which may be accepted in order that the relationship between variables may be tested empirically. For example, the following are both hypotheses; that vegetation patterns change with geology and soil, and that pedestrian flows are at their highest near the peak land value intersection.

The hypothesis will state the nature of the relationship between the variables and may be expressed as a *null* hypothesis stating that there is *no* relationship between the variables, the reverse of the expected. The *research* hypothesis states that there *is* a relationship and the hypotheses given above are strictly of this type.

ICE AGE

The common term given to a period of glacial advance and activity such as the last glacial period which occurred from about 2 million years ago until approximately 10,000 years ago and is known as the Pleistocene period. The Ice Age had a series of **interglacial** periods within it when the ice retreated and this may have occurred up to 20 times.

ICE CAP

◀ Ice sheet ▶

ICE FIELD

◀ Ice sheet ▶

ICE LENSES

A horizontal bed of ice in the ground which grows very much like an **ice wedge** and leads to the formation of **hummocky ground** and **pingos**.

ICE SHEET

A more extensive form of **glacier** covering a larger area such as Antartica. It is sometimes known as an **ice cap** when it covers a slightly smaller area. An **ice field** is a smaller area such as the Columbia Ice Field which feeds a series of glaciers in Alberta.

ICE WEDGES

A crack in the ground which forms in areas of **permafrost** and develops from a smaller ice crack. It fills with water in the summer, which then freezes and expands. In the following summer the larger crack refills with water and the process is repeated, enabling an ice wedge to reach a diameter of over 9 metres.

IDEOLOGY

This term can be used in several ways but it generally tends to refer to the system of ideas and beliefs which relate to a particular view of reality which a group may hold. Therefore, 'Thatcherism', for example, is seen to have a particular ideology based on freedom of the individual, monetarism and rolling back state control which then influences the interpretation of all other events which take place. It is often used to show the ideas and beliefs of a particular social or cultural group.

IGNEOUS ROCK

Rock formed by solidifying volcanic magma either inside the earth as **intrusive** landforms or on the earth's surface as **extrusive** landforms. The composition of the rocks varies according to the chemical composition and the speed of cooling. Metamorphic rocks are sedimentary rocks whose composition has been altered by proximity to the heat of **igneous rocks**.

ILLUVIATION

The process of deposition of soil material removed from one horizon to another horizon in the soil. It produces *illuvial horizons*.

IMAGE

1 The term, used especially in aerial photography and **remote sensing**, which refers to the photograph which has been taken of an area of the earth's surface.
2 The mental recall of a map, picture, person or event which may be described verbally or drawn (in the case of a map or diagram) is a **mental map**.

IMMIGRATION

The movement of people into an area which normally results in a growth of the population. Immigrant groups are generally younger than the host population and are therefore more fertile. The fertility of immigrant groups often leads to political unrest in the host population (e.g. in West Germany the extreme right of the political spectrum opposes labour migration because of the high **birth rate** among **labour migrants**).

IMPERMEABLE

A term which refers to the fact that a rock will not permit water to pass through it due to the lack of pore spaces or the lack of joints and bedding planes.

IMPORT SUBSTITUTION

The replacement of costly imports by locally made produce. It enables the economy to diversify and also enables the workforce to acquire new skills. Developing countries will gain most if there is some local equity in such developments. In Malaysia the Mitsubishi car plant was developed primarily to provide the country with its own car industry and now markets its cars as Protons, a Malaysian car, so saving on costly automobile imports.

INCISED MEANDER

◀ Ingrown meander, Intrenched meander, Meanders ▶

INDEPENDENT VARIABLE

◀ Dependent variable ▶

INDUSTRIAL DEVELOPMENT CERTIFICATE (IDC)

The document required to obtain permission to build a factory of a certain size within the areas of control defined by acts of parliament starting with the Distribution of Industry Act 1945 until the demise of regional policy in the 1980s. IDCs were not required by firms wishing to build a plant in the special development areas, development areas and intermediate areas.

INDUSTRIAL GEOGRAPHY

A branch of economic geography which examines the secondary sector, its patterns of, and changes in, location.

INDUSTRIAL INERTIA

The tendency for an industry to remain in an area despite a more favourable location existing elsewhere. The main deterrent to a move is the value of the capital invested in the site; this partially explains the continued existence of the coalfield steelworks in Duisburg when coastal locations would be more profitable. It is sometimes called geographical inertia when it is applied to other activities.

INDUSTRIALISATION

A process in which industrial employment comes to dominate the employment in a country over a period of years and industrial products form the basis of the economy. In the past the process was one which emerged from cottage industries serving local markets and mercantile trading with colonies as the 'industrial revolution'. The new industrial system was based on factory work and the division of labour within the factory and between the factory and home. In other parts of the world industrialisation became a part of the planned development of the society, e.g. East Europe after 1945 when further industrialisation was the subject of national plans. Industrialisation is also a strategy of many Developing countries who seek to achieve newly industrialising country status or at least a planned step along the path to greater industrialisation.

INDUSTRIAL LOCATION THEORY

A set of theories which attempt to explain the reasons for the location of manufacturing plants. It depends on the work of Weber, Lösch, Hotelling, Hoover and Smith. These theories have been replaced by others which depend on interpreting behaviour, and the nature and structure of the firm.

INFANT MORTALITY RATE

This is an age specific death rate which measures the number of deaths to children aged under one year per thousand live births. It is an important indicator of health care and an indicator of the level of social development in a country. In 1985 the rate in Nigeria was 105/1000, Bangladesh 133/1000, Brazil 71/1000, United Kingdom 10.1/1000 and Sweden 7/1000.
◀ Neo-natal mortality rate ▶

INFILTRATION

The process in the hydrological cycle by which precipitation enters the soil where it can be stored or from whence it can percolate to the groundwater store or move by throughflow to the channel store. The speed at which it enters the soil is known as the infiltration rate and may be measured using an infiltrometer. If the intensity of the precipitation exceeded the speed at which the water can infiltrate the soil then the infiltration capacity is reached and overland flow commences.

INFILTRATION CAPACITY

The maximum rate at which water can enter the soil under specifically defined conditions with respect to moisture content.

INFILTRATION RATE

The actual rate at which water is entering the soil at any given time.
◄ Infiltration ►

INFILTROMETER

A device for measuring the rate of entry of fluid into a porous body, e.g. water into the soil.
◄ Infiltration ►

INFLATION

An economic term which refers to the process of rising prices which reduces the value of money. Very often this may become an inflationary spiral where the rising prices continually trigger demands for higher wages which in turn fuel inflation. If the demand for goods and services runs ahead of the ability of the economy to produce the goods to pay for that demand, inflation will result. Politicians disagree about the best way to cure inflation.

INFRASTRUCTURE

Those facilities which provide the basic framework of an economy which include transport in all its forms, communications, power supplies, water supply and waste disposal.

INGROWN MEANDER

A form of incised meander where the meanders have gradually incised as a consequence of slow rejuvenation allowing the meanders to develop faster than the downcutting. This produces an assymetrical valley with a steep meander scar or bluff on one side and a more gentle slip-off slope on the other.

INNER CITY PROBLEM

A very complex issue which has put together many problems of society which are found most in inner city areas and suggested that they are linked by location. The problem involves issues concerning **deindustrialisation**, population migration, building decay and redevelopment, poor infrastructure, unemployment and abandoned land. The policies to solve the problem include **partnership authorities, programme authorities, designated districts, garden festivals, enterprise zones,** urban development corporations **city action teams** and urban renewal grants. Some of the results of policies are **gentrification,** new industrial zones and waterfront developments. The social structure of many areas is changing and many see this as further disadvantaging the poor.

INNER URBAN AREAS ACT 1978

This set up **partnership areas, programme authorities** and **designated districts.**

INSOLATION WEATHERING

The shattering of the outer layers of rocks as a result of the diurnal fluctuation of air temperature raising and lowering the temperature of the rock.

INTEGRATED DEVELOPMENT OPERATION (IDO)

A new (1988) **European Community** initiative to **target** assistance to specific places rather than permit whole areas to quality for help. Belfast and Birmingham together with Naples were the first cities to obtain this regional aid.

INTENSIVE AGRICULTURE

Agriculture which has high levels of capital, labour and fertiliser inputs and high levels of output for each hectare of land. The most intensive forms of agriculture are **market gardening, horticulture** and **glasshouse crops.** Freezing and canning have also resulted in the intensive farming of vegetable crops in close proximity to the large freezing factories which initially developed to freeze fish, e.g. Lowestoft. Improved transport has enabled intensive farming to move further from the market as has happened with truck farming in the USA. Today goods can be transported in refrigerated trucks or even by air to the market. Other intensive farming types are pick your own (PYO) farms, battery poultry and egg farms, and intensive dairying and meat production. **Irrigation agriculture** is another type of intensive farming.

INTENSIVE SUBSISTENCE SEDENTARY AGRICULTURE

A tropical agricultural system associated with SE Asia. It is based on the necessity to provide food for the large population of the region and the fact that the green revolution has enabled a surplus of rice to be grown which is exported.

INTERCEPTION

The role played by plants in preventing precipitation from reaching the ground surface.

INTERCEPTION STORAGE

A stage in the hydrological cycle in which precipitation is held in the vegetation either to evaporate or to fall to the lower vegetation and the ground by stemflow and drip.

INTERDEPENDENCE

The matrix of ties which bind a society or an economy together are interdependent. People perform roles as an individual, a member of a household, neighbourhood, workforce and an electorate and these roles all interlock with those of other citizens. Similarly a company is interdependent with suppliers and purchasers.

INTERFACE

The boundary between two systems which is not clearcut, e.g. the interface between recreation and tourist needs in a resort. It can also refer to the boundaries between disciplines or areas of a discipline, e.g. industrial location can be seen as being at the interface between economic geography and economics and even other disciplines such as psychology and sociology.

INTERGLACIAL

◀ Ice age ▶

INTERMEDIATE AREAS

The Hunt Committee proposed these areas which were implemented in 1970. They are also known as 'grey areas'. These areas had economic difficulties, though less than those of the **development areas**. Selective aid for industry was available. In 1987 these areas and the development areas were abolished.

◄ Development areas ►

INTERMEDIATE TECHNOLOGY

Schumacher developed the idea that both capital- and labour-intensive technologies are inappropriate for the developing world's economic advance. He suggested that intermediate technology would bridge the gap between the traditional technologies of the developing world and the new technologies and enable development to be more suited to the needs of the developing countries. Rope wells and simple irrigation systems funded by the charity Bandaid are one example of intermediate technology in practice. The promotion of local craft skills for tourist souvenirs is another example of the same theory in action. It must be remembered that intermediate technology is unlikely to bring rapid development despite the technology being more appropriate for the countries' needs. The term appropriate technology is also used for these developments.

INTERNAL DEFORMATION

◄ Basal sliding ►

INTERNATIONAL COMMODITY AGREEMENT

A trading commodity agreement which is signed by countries to try to maintain the price of a commodity so that the stable price will benefit the suppliers in particular. It is an attempt to end the wild fluctuations in price of commodities such as coffee, rubber and cane sugar. On the whole the schemes have had only limited success because they have not really controlled the fluctuations in price which both make and ruin producers.

INTERNATIONAL DIVISION OF LABOUR

The division of economic activity which increasingly results in the control of economic production together with product innovation in the **developed world** and the routine production of goods in the **developing world** where the labour costs in particular are cheap. It is largely the result of the strategies of the **transnational corporations** and has led to the development of the **newly industrialising countries** whose economic advance has been founded upon producing goods for the transnational corporations. This is all part of the process of **global shift**.

INTERNATIONAL MIGRATION

The movement of people between countries either on a permanent basis, e.g. migration to the USA in the nineteenth century or as **circulatory migrants**, e.g. **labour migrants** to West Germany between 1960 and 1990. The reason for the movement is based on the perceived advantages and disadvantages of the place of origin and destination together with consideration of **intervening obstacles** such as the cost of migrating and the distance involved. Very frequently the presence of family in the place of destination to aid the process of adjustment is a major pull factor. The causes are:

population pressure and inadequate resources;
shared poverty;
better employment prospects;
political factors;
religious persecution;
racial prejudice;
better value placed on skills at destination;
ties between states existing from colonial times;
freedom of labour movement.

INTER-REGIONAL MIGRATION

This takes place within a country and can be from rural to urban areas, i.e. **rural depopulation**, and from urban to rural areas, i.e. **counter-urbanisation** movements such as from the Scottish Highlands to Central Scotland or Yorkshire to the South East. Generally such moves are associated with work opportunities although other factors such as promotion in a national organisation (e.g. banks), marriage and remaining in an area upon graduation also feature as reasons for migration. As in all migration the perceived advantages and disadvantages of the places of origin and destination together with intervening opportunities affect the decision to migrate. In the **developing world** inter-regional migration will generally be the primate city or the major urban areas of the country and more in line with **Ravenstein's laws of migration**.

INTER-REGIONAL TRADE MULTIPLIER

A complex form of analysis which seeks to establish the transactions between regions and between sectors of the economy within regions. It is more sophisticated than **export base theory** in that it measures inputs into a region or company and outputs from that region or company in a complex analysis which shows how the transactions of a region build up into its economy. It can be used predictively to estimate the effect of, e.g. the opening of the Toyota plant on the economy of the East Midlands.

INTER-TROPICAL CONVERGENCE ZONE (ITCZ)

The equatorial zone between the northern and southern Hadley cells where the converging trade winds are forced to rise in the equatorial trough of low pressure. This results in great instability, producing heavy convectional rainfall. During the northern summer the ITCZ is displaced northwards drawing the trade wind belt northwards and bringing seasonal precipitation to the savanna biome. It is drawn further north (and south) over the land masses. The distance which the ITCZ moves north and the timing of its movements are a critical element in the agricultural calendar in the savanna and monsoon lands. The non-arrival of the associated rains is one cause of desertification. Why it fails to move on a regular basis is not clear although some link its movement to the changing effects of solar radiation on the currents in the Pacific Ocean and the so-called El Nino effect.

INTERVAL DATA

Data which indicates the value along a continuous scale where the intervals are fixed. The units of measurement can take almost any form from paces to metres or degrees celsius.

INTERVENING OBSTACLES

◀ Stouffer's intervening opportunities ▶

INTERVENTION PRICE

A form of subsidy which sets minimum prices for crops to ensure a fair standard of living for farmers. It is a major component of the Common Agricultural Policy of the European Community in the form of the European Agricultural Guidance and Guarantee Fund (EAGGF). If the market price falls below the intervention price, the crop is bought at the intervention price by the EC. The system leads to overproduction and leaves large stocks of surplus products in store such as 'butter mountains' and 'wine lakes'. Marginal area farmers are encouraged to grow produce by the guarantee of a basic income and so crops may be grown in areas which are not best suited for them.

INTRA-REGIONAL MIGRATION

Movement within an area generally associated with changed residence and the life cycle. It tends to be i) a movement towards the suburbs, i.e. suburbanisation, ii) counter-urbanisation movement, or iii) moves back into a city such as those associated with re-urbanisation and gentrification.

Intra-regional migration in the developing world will be into the city and in contrast to that in the developed world.

INTRA-URBAN MIGRATION

◀ Filtering, Life cycle residential moves ▶

INTRENCHED MEANDER

A form of incised meander which has developed in an area where rejuvenation was rapid so that the vertical erosion of the channel was more pronounced than lateral erosion thus leaving the meanders in a steep-sided valley.

INTRUSIVE VULCANICITY

Magma is forced into the country rock but does not reach the earth's surface. Subsequent erosion will often expose the intrusions on the surface.
◀ Batholiths, Dykes, Laccoliths, Sills ▶

INVASION

Burgess used the term to imply the movement into residential areas of new social groups who would then come to dominate the community in the same way that plants in the biosphere come to dominate an area by invasion and succession.

INVISIBLE EARNINGS

The earnings of a country obtained through invisible exports. Services such as insurance are sold to foreign countries and the payment for the service, unlike normal exports, is not in return for goods. Tourist earnings are also an invisible export but our travel abroad is an invisible loss to the UK economy.

INWARD INVESTMENT ORGANISATIONS

Private capital initiative to promote regional development, e.g. Northern Development Company.

IRRIGATION

The application of water to water-short areas which are not necessarily arid, to increase the output of these areas. In some areas the natural flooding of rivers has been utilised as a form of irrigation for padi rice, while in other areas people have invented increasingly sophisticated methods of moving water from the rivers and underground to the fields. Simple irrigation systems which involve gravity flow and low energy devices to raise water are characteristic of the **developing world**. The more complex systems using barrages to store water, deep wells and electric pumps are a greater danger to the soil if the technology is not carefully controlled. **Salinisation** may take place due to evaporation.

IRRIGATION AGRICULTURE

A form of **intensive agriculture** which depends upon the controlled, artificial distribution of water to the crops to assist plant growth. It was associated with areas which are arid and semi-arid although irrigation is used today to assist crop growth if precipitation fails to reach the optimum for crop growth during the drier months. In Britain it has been increasingly used during dry summers to aid the growth of field vegetables, maize, and even pasture. Other areas of W. Europe depend very heavily on irrigation such as the valley lands around the Mediterranean, e.g. Huerta of Spain and areas of intensive crop production such as the Rhine Rift valley and the Lower Rhone. The semi-arid and arid areas of the world either depend on sophisticated new methods of irrigation based on deep wells and abstracted river water, or on the centuries-old methods of flood irrigation, small dams and barrages, and the labour and animal intensive methods of raising water from wells and across river banks.

ISLAND ARCS

These occur where two oceanic plates collide along a **destructive margin**. The islands of the West Indies such as St Lucia and Martinique are part of this arc of volcanic activity associated with this **subduction zone**.

ISODAPANES

Lines of equal total transport cost derived from **isotims** which are cost contours around a least cost area in the **Weber model of industrial location**. It is assumed that they are drawn on an **isotropic surface**. See Fig. I.1.

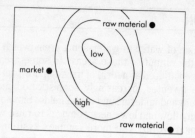

Fig. I.1 Isodapanes

ISOLINE MAP

A map which is like a contour map in that it joins together lines of equal value of the variable which is being measured on a **meteorological** chart are a form of isoline as are isotherms. The map in Fig. I.2 shows isolines of pedestrian counts in Blackburn.

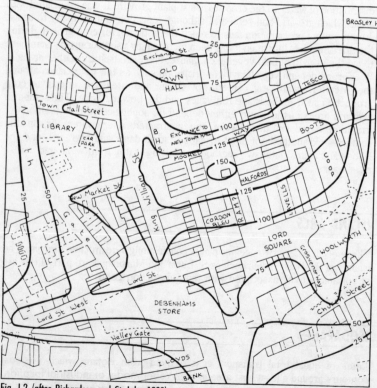

Fig. I.2 (after Richardson and St John 1989)

ISOSTACY

The balance which exists in the earth's crust between the weight of the larger, high landmasses and a root of material descending into the mantle to maintain balance. This balance may be affected by the weight of glaciers which push the crustal plates deeper into the mantle so that, when the ice melts, the land mass rises again as a result of the process of **isostatic readjusment**.

ISOSTATIC ADJUSTMENT

Changes in the level of the oceans which are due to a movement of the land. The land may move upward due to the lessening of the weight of the land as a result of erosion or downwards due to increased weight from glacial ice. It is isostatic adjustment which is causing southern England to sink in compensation for the rise of the land further north since the retreat of the Pleistocene ice caps.

ISOTIMS

Lines of equal cost of transporting raw materials and the finished product as shown in Fig. I.3. They are used in the construction of the least cost in **Weber's model of industrial location**. It is assumed that they are drawn on an isotropic surface.

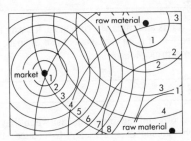

Fig. I.3 Isotims

ISOTROPIC SURFACE

A surface which assumes the same physical characteristics in all directions. It is assumed to be featureless with the same soil characteristics, population density and potential for movement. The surface is an essential component of the theories of **Christaller, Von Thünen** and **Weber**.

JAPANISATION

The consequences of the dominance of Japanese investment and the adoption of Japanese methods of production and management for economies outside Japan. Japanese methods are introduced in the branch plants of Japanese companies in the host country and are then adopted by other firms in that region or that industry.

JOURNEY TO WORK

The travel from home to the place of work and back which is the main demand for public transport. The area from which people commute to work is called the 'journey to work region' and is frequently used in economic and physical planning as a means of delimiting the urban field or hinterland of a city.

JUST IN CASE

The delivery of goods and raw materials in such a way that they can be stored for use when needed. It is an extravagent use of space and has been replaced by the just in time principle.

JUST IN TIME

The delivery of goods and raw materials just when they are needed. This was developed by the Japanese as a means to reduce storage and warehouse capacity. It is an aspect of Japanisation.

KAME

A **fluvio-glacial** feature built from materials formed when a **crevasse** in a glacier has filled with debris. The ice melts and leaves behind a conical hill of bedded material. Kame terraces form along the edge of a glacier and a valley and are composed of material deposited by **meltwater** streams flowing alongside the glacier.

KARST

The name of a region in Yugoslavia which has given its name to the scenery associated with **carboniferous limestone**. Karst may be found in other limestones. In this landscape **chemical weathering**, **carbonation** and fluvial **erosion** dominate. Carboniferous limestone is composed of massive blocks whose joints influence the processes of weathering and erosion.

KATABATIC WIND

A downslope wind in mountain valleys caused by colder air being drawn downslope and sinking downslope as the air cools, as shown in Fig. K.1.

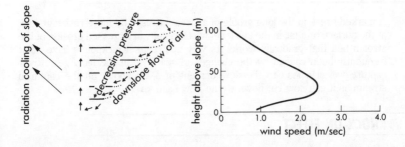

Fig. K.1 Katabatic airflow

KETTLE HOLE

A depression in an area of glacial deposition formed when a block of buried ice melts and the surface collapses and fills with water. Most kettle holes become filled with sediments which are deposited in them by water.

KEY SETTLEMENT

A term used interchangeably with *key village* in rural planning to signify a village which has been chosen to receive the lion's share of investment in services and housing in order that a **threshold** population will be reached and services further improved.

KIBBUTZ

A form of **collective farm** found in Israel.

KINETIC ENERGY

◀ Energy ▶

KING'S PARALLEL SLOPE RETREAT MODEL

A model based on observations in semi-arid areas when the **cliff** and **debris slopes** remain the same proportion of the total slope length; and retreat in parallel leaving an increasingly large concave lower slope or **pediment**. The final stage is the formation of a **pediplain**. This model is best applied to the mesa and butte landscapes of North America.

KNICKPOINT

A marked break in the long profile of a stream caused by more resistant rock at the surface or a fall in the sea level which results in increased erosion of the stream bed that gradually works its way upstream. Knickpoints have most frequently been caused by the changes in sea level in the Ice Ages. River capture will also lead to the development of a knickpoint on the captured stream as it can now cut down to the new base level.

KNOCK-ON EFFECT

A term which is generally synonymous with **multiplier effect**. It refers to the effects that one decision may have on another such as the effect on its suppliers of the decision to relocate a factory. Natural events may also give rise to knock-on effects, e.g. landslip which then results in the damming of a stream which may cause it to dry up below the temporary dam.

KONDRATIEFF WAVES

Sometimes referred to as cycles or long waves. Kondratieff noted that the world economy developed in a wave-like progression with each upturn being the result of a set of new technologies. The first wave was the product of the Industrial Revolution between the 1780s and 1840s when economic growth was based on steam power. The second wave between the 1840s and 1890s was dominated by the technology of steel. The third wave occurred between the 1890s and the 1930s and was dominated by electric power, the automobile industry and chemicals. The fourth wave after 1940 has been dominated by the aeroplane, electronics and computing. Some talk of a fifth Kondratieff which began in the 1980s and is based upon the microchip technologies and biotechnologies.

KONIG NUMBER

A measure of accessibility used in **network theory** which calculates the maximum number of edges from any one vertex by the shortest path to any other vertex in the network. The lower the value the greater the centrality.

LABOUR COSTS

The total costs of wages and salaries and all associated payments such as National Insurance and pension contributions and including perks such as company cars and health insurance. Labour costs are a significant element of most economic activities and reflect the skills of the job and scarcity of that skill. Labour costs for women are still lower partly because many women are employed part time where the associated costs are lower. This is one reason why the female labour force has grown in recent years. Labour costs are generally lower in the **developing world** and this factor alone accounts for the shift of much manufacturing overseas and the **deindustrialisation** of manufacturing regions. The result of the shift to the low wage economies is often termed the **international division of labour**.

LABOUR FORCE

◄ Active population ►

LABOUR MIGRATION

A form of **circulatory migration** drawn by labour shortages in advanced industrial economies from less advanced economies which either have had a colonial link, e.g. West Indians to London in the 1950s, or are a relatively short distance away, e.g. Turkey and Greece to West Germany. Labour is normally brought on contract for specific periods but many have now stayed in their adopted country to be joined by their families. Economic union and the free flow of labour in the **European Community** has made it easier for labour to move between the member states although movement from outside of the community is more rigidly restricted and schemes exist to encourage **return migration**. Another variant is the skilled labour migrants working for major international companies who can be moved to specific posts such as working for the oil-rich states, or computer and medical experts drawn to the high wage economies of, e.g. Paris or Dallas. The popular term for this form of migration is **brain drain**.

LACCOLITH

An intrusive dome of magma beneath the surface which forces the overlying sediments upwards. Subsequent erosion of the overlying sediments may expose the solidified magma. Unlike the larger batholith it is fed from a pipe leading to the magma chamber.

LAGGING REGIONS

A term used by Myrdal to describe those regions or countries whose economy was declining due to the backwash of labour, capital and resources to the core region.
◄ Myrdal's model of cumulative causation ►

LAG TIME

◄ Hydrograph ►

LAND CONSOLIDATION

An aspect of the European Community's Common Agricultural Policy which aims to create compact landholdings around a farm in those areas where inheritance has resulted in scattered landholdings and farm fragmentation. It is often known by its French term *rembrement* although this government-backed policy may be found in West Germany and the Netherlands and, outside the EC, in Austria and Switzerland. The new compact holdings enable capital inputs to increase and labour inputs to decrease leading to an increased output. The building of new farms away from the village and the consequent dispersal of population, and the building of new roads also form a part of the policy.

LAND REFORM

◄ Enclosure ►

LAND-SEA BREEZES

Because land heats faster than water (which needs five times the energy to raise its temperature by a similar amount) the air over the land will expand as a result of radiation from the sun, rise and thus draw in air from over the sea. This air is cooler and moist and will often be foggy especially in the spring. The reverse pattern happens at night. Lakes may also cause a similar small scale wind pattern on a diurnal basis. It develops best during anticyclonic conditions.

LAPILLI

◀ Volcano ▶

LAPSE RATES

DALR

The DALR (dry adiabatic lapse rate) is the rate of fall of temperature of air rising adiabatically up through the atmosphere. It has a value of 9.8°C per km.

▶ SALR

The SALR (saturated adiabatic lapse rate) is the rate of fall of temperature of saturated air which is rising up through the atmosphere and whose temperature has fallen below the dew point. It is less than the DALR because as the air is cooled, water vapour condenses out of the atmosphere, releasing the latent heat of vaporisation. The difference between the SALR and DALR therefore depends on the amount of water vapour condensed out of the air. As warm air can hold considerably more water vapour than cold air, the greatest difference between the SALR and DALR is found at low levels in the tropics and subtropics. The least difference between the SALR and DALR is found at high latitudes in winter, when temperatures are very low, and also in the upper levels of the atmosphere where temperatures are similarly very low. In fact, once the air temperature is below about −40°C, there is so little moisture in the air that the amount of latent heat released is so small that the SALR is nearly the same as the DALR. The SALR therefore has a range of values depending on temperature from about 3.5°C per km in hot tropical air to over 9.5°C per km in cold air where the temperature is below −40°C.

◀ Stability ▶

LATE DECLINING STAGE

◀ Demographic transition model ▶

LATE EXPANDING STAGE

◀ Demographic transition model ▶

LATERAL MORAINE

◀ Moraine ▶

LATE TRANSITIONAL SOCIETY

◀ Mobility transition model ▶

LATOSOL

A soil type used in the old zonal classification schemes. Soils under forest in tropical, humid conditions. Red coloured, low base status, low in most primary minerals, and high in low activity 1:1 lattice clays.

LAVA

Molten **magma** from the earth's crust which has reached the surface following volcanic activity.

LAVA FLOW

A stream of lava which flows from a vent or fissure. If the flow is of basic lava it will spread further as, for example, in the Deccan areas of India. It is a form of **extrusive vulcanicity**.

LEACHING

The removal of materials in solution from the soil. It is part of the process of **eluviation**.

LEADING INDUSTRY

◀ Growth pole theory ▶

LEAST DEVELOPED COUNTRY (LDC)

◀ Fourth world ▶

LEE'S LAWS OF MIGRATION

Four groups of factors underlie the decision to migrate;
    factors linked to the destination of the migrant;
    factors associated with the area of origin of the migrant;
    intervening obstacles between origin and destination;
    personal factors.

LEISURE

A broad term which is taken to refer to those activities which are undertaken outside of the working day/week. It is that time where the choice of activity is up to the individual. It can refer to **recreation, tourism,** socialising with friends, playing sport or attending cultural events. Leisure may be home-based or non-home based and it is this latter form of leisure which has been the subject of geographical study. Other activities are now being promoted as leisure and many new developments in retailing (in particular such as Gateshead Metro-centre and West Edmonton Mall) have been developed around a 'leisure and shopping' theme.

LESS DEVELOPED COUNTRY

◀ Developing world ▶

LESSIVAGE

A French term describing the process by which the fine clay is translocated in suspension by the soil water from the eluvial horizon and deposited in the form of clay skins (cutans, argillans) in an illuvial (B_t) horizon.
◀ Illuviation ▶

LEVEE

The raised bank which develops along a river channel formed by the stream flooding over the banks and depositing its coarser, heavier material near to the river channel. It results in the river channel being higher than the surrounding floodplain. People have tried to reduce the risk of flooding by building **artificial levees.**

LIFE CYCLE RESIDENTIAL MOVES

These are best described by Figs. L.1 and L.2.

LIFE EXPECTANCY

The average number of years which a person can expect to live. It is normally expressed as years from birth. In 1985 it was 50 years in Nigeria, 48 in Bangladesh, 63 in Brazil, 73 in the United Kingdom and 76 in Sweden. It is normally expressed as separate figures for males and females and it can be expressed as the number of years of life expected by each surviving person at each age. Life expectancy has risen due to improvements in health care which initially increased longevity but, more recently, due to improved care of the newly-born child.

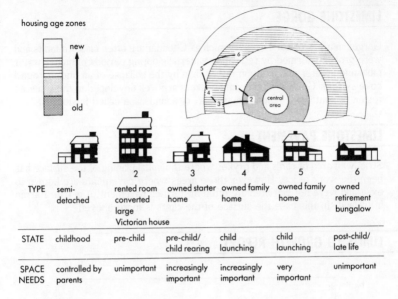

housing age zones

new

old

	1	2	3	4	5	6
TYPE	semi-detached	rented room converted large Victorian house	owned starter home	owned family home	owned family home	owned retirement bungalow
STATE	childhood	pre-child	pre-child/child rearing	child launching	child launching	post-child/late life
SPACE NEEDS	controlled by parents	unimportant	increasingly important	increasingly important	very important	unimportant

Fig. L.1 Life-cycle moves in a British city

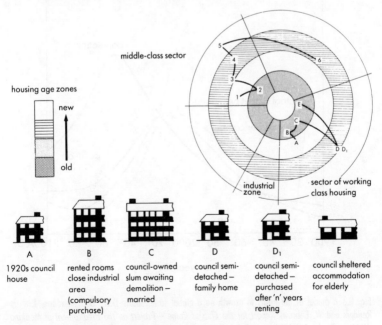

middle-class sector

housing age zones

new

old

industrial zone

sector of working class housing

A	B	C	D	D₁	E
1920s council house	rented rooms close industrial area (compulsory purchase)	council-owned slum awaiting demolition – married	council semi-detached – family home	council semi-detached – purchased after 'n' years renting	council sheltered accommodation for elderly

Fig. L.2 Contrasting life-cycle and social housing moves in a British city

LIMESTONE GORGE

Gorges in areas of **karst** scenery such as Cheddar are often close to faults and were probably formed by retreating waterfalls during periods of higher water table and greater precipitation rather than by the collapse of an **underground cave system**. Gorges which contain rivers are well developed in the Causses region of southern France and are part of a landscape called **fluviokarst**.

LIMESTONE PAVEMENT

Strictly an area of **clints** and **grykes** where the initial planing of the surface has been the work of ice. However those areas where pavements are known to exist and were not glaciated are often mistermed pavement, e.g. The Burran in Ireland. In this case the surface of the clints is more uneven.

LIMITS TO GROWTH REPORT

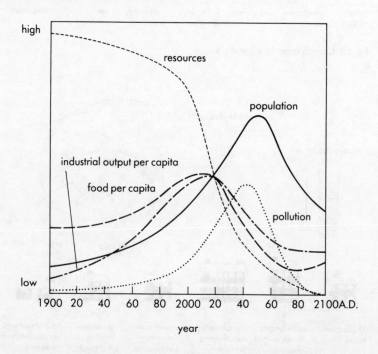

Fig. L.3 A model of the limits to growth on a global scale. *source* D.H., D.C. Meadows, J. Renders and W. Behrews *Report for the Club of Rome – Project on the Predicament of Mankind* (Universe Books, NY 1972)

This neo-**Malthusian** report was published by the **Club of Rome** in 1972. The group suggested that population growth is determined by the interaction of five factors; population, food production, natural resources, industrial production and pollution. They concluded that if all of these factors continued to grow exponentially then the capacity of the earth to sustain growth would be reached by the year 2070. Then both population and industrial capacity would decline. They suggested that this scenario could be avoided by policies of economic and ecological stability which would bring about a state of **global equilibrium**, a balance between population and the resources needed to support that population, see Fig. L.3.

LINE GRAPH

The most common method to display results which can take account of temporal variations where they are most useful. Fig. L.4 is an example.

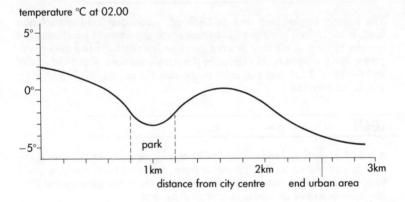

Fig. L.4 A line graph

LIQUIEFACTION

◄ Earthquake ►

LITHOSOL

Often called intrazonal soil, it is a soil whose formation is not related to the major climatic zones of the world but to rocktype or the presence of water. **Gley** soils are a good example of such a soil as are saline soils in marshland areas along the coast.

LITHOSPHERE

The rocks of the earth's crust which make up the seven major and twelve minor plates. Together with the **hydrosphere** and **atmosphere** it is one of the three major constituents of the earth's environmental system. The lithosphere is the location for the forces of **weathering** and **erosion**. It also provides nutrients for the **biosphere**.

LITTER LAYER

The layer of leaves and plant debris which covers the soil and which provides the humus. It is a rapidly decaying layer in the **tropical rain forest** and a slowly decaying layer in the **boreal forest**.

LOAD

The material carried by a river as **bedload, suspended load** and **solution load**. It has arrived in the river as a result of the processes of weathering and erosion. Some load will have entered the river as a result of **bank caving**, the cutting into the banks of the channel by the stream and the collapse of the bank into the river. Load may also refer to the material carried by glaciers or by waves and currents.

LOAM

A soil which combines the characteristics of both sandy and clayey soils in the most favourable way so that it is well aerated, drains relatively freely, is easily tilled and is fertile. Loams can be dominated by one constituent so that they are often described as sandy, silty or clayey loams.

LOCAL GOVERNMENT AND PLANNING ACT 1980

The legislation which governs historic buildings and conservation areas.

LOCAL PLAN

◀ Town planning ▶

LOCATION DECISION MAKING

An attempt to make us aware of the whole decision making process in which organisations operate which has developed from the work of **Weber, Lösch**,

Smith, Pred, Hotelling and others. It may be used for the locational study of an industry, a shop or any other service activity and is illustrated in Fig. L.5.

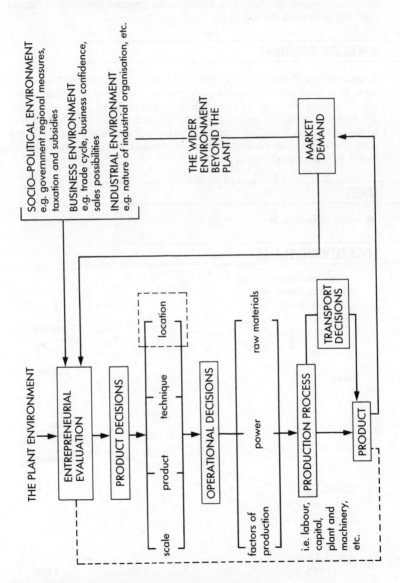

Fig. L.5 Location decisions as part of the total investment process (based on Gold)

LOCATION OF OFFICES BUREAU (LOB)

The body established to advise office firms to relocate away from London and to try and encourage offices to move to the **assisted areas**. The bureau operated between 1964 and 1979.

LOCATION QUOTIENT

A statistic which measures the degree of concentration of an activity in an area. It is used most in studies of **agglomeration** to identify those areas where a particular type of employment is concentrated. The quotient is obtained by dividing the percentage of the population in an activity in an area by the percentage found in that activity in a broader region or a country. If the quotient exceeds 1 that activity is more concentrated in that area; if it is less than 1 it is more dispersed in that area.

LOESS

◀ Wind deflation ▶

LOGARITHMIC SCALES

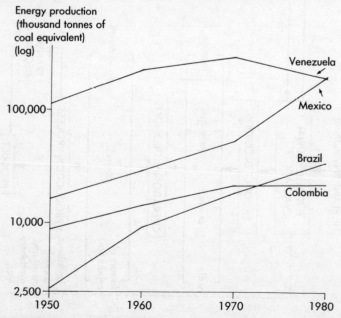

Fig. L.6 Total production of energy (in thousand tonnes of coal equivalent)

A scale such as the **Richter scale** where the numerical difference is not constant as on an arithmetic scale, but proportional. It is useful for plotting data which has a great range of values (see Fig. L.6) and when such a scale is plotted against time. Any increase which is doubling over each period of time will appear as a straight line on semi-logarithmic paper whereas it would be a steepening curve on arithmetic paper.

LOGISTIC CURVE

Sometimes called an S-shaped curve. It has been used in population studies to show the nature of population growth over time. There is accelerating growth until the carrying capacity of an area is neared. Then the growth slows and enters a steady state when the population level stabilises.

LONG PROFILE

The graphical representation of the height of a river above sea level from its source to its mouth or to its confluence with another stream. Study of the curve will indicate **knickpoints**. The nature of the curve may be expressed mathematically. The general shape of the profile or profiles is concave because the river has greater discharge which enables it to transport the load, the **channel** is more efficient and the smaller particles can be more easily transported over a gentle slope.

LONGSHORE DRIFT

A process which results from the fact that **waves** rarely approach the shoreline at right angles. The angle of approach ensures that the particles in the **beach store** are moved up the beach at an angle by the **swash** but move at right angles down the beach with the backwash. Particles are therefore moved slowly along the beach in the predominant wave direction; this process speeds up under storm conditions. It is sometimes called littoral drift.
◀ Groynes ▶

LORENZ CURVE

This illustrates the degree of unevenness in the distribution of population or production or income. An even distribution is represented by a straight line across the graph at 45° but most distributions are not even, and the curve lies below the straight line.

LÖSCH

Losch developed the **central place model** of **Christaller** to include **city rich city poor** sectors. He also noted that it was not necessary for all central places to include functions of a lower order centre as well as those of their own order. This enabled the model to be refined to be closer to reality.

LÖSCH'S THEORY OF INDUSTRIAL LOCATION 1954

A theory of location which emphasises the role of demand from the market. The theory illustrated in Fig L.7 maximises demand at the market with

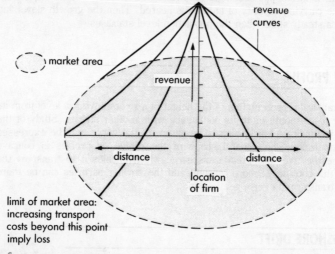

Fig. L.7 Lösch's revenue cone

revenue falling as plants are located further from the market. The revenue curve around each market will then form a cone delimiting the market area where revenue receipts will exceed costs. **Agglomeration** should take place around the point of maximum revenue. In some circumstances dispersal will result as each company manufacturing a product seeks out its own market. Lösch, like **Weber**, assumed perfect competition, rational economic people and an **isotropic surface**. His theory ignores the difficulties of the decision makers who may not know everything about the revenue of a firm. It was subsequently modified by others although its principle of market orientation does enable it to be applied to industries such as brewing.

LOW STATIONARY STAGE

◀ Demographic transition model ▶

MALTHUS'S THEORY OF POPULATION GROWTH

Malthus wrote 'An essay on the principle of population' in 1798 in which he put forward the view that population grows faster than the means of feeding that population. Population, if unchecked would grow at a geometric rate, i.e. 1,2,4,8,16, whereas subsistence (not merely food supply) would grow at an arithmetic rate, i.e. 1,2,3,4,5. He calculated that in 200 years the ratio of population to subsistence would be 256 to 9 and eventually checks on the population's growth would occur. **Positive checks** include war, disease, poverty and a lack of food and would increase the **death rate** and so reduce population pressure. **Preventative checks** were many forms of moral restraint such as the postponement of marriage, and vice by which he meant adultery, birth control and abortion which would affect the **birth rate**. The theory has not come to fruition as rapidly as Malthus predicted although famines in the **developing world** do suggest that positive checks are becoming important today. The slow realisation of his predictions in the **developed world** can be explained by;

the fact that he wrote before the Industrial Revolution;

he did not foresee the growth in agricultural output in the new world, from improved crop strains and new crops;

the impact of improved transport of food, e.g. refrigeration;

the medical advances in birth control.

Also he confused moral and religious issues with population issues. Malthusian ideas have been criticised because the predictions have not been realised and because some believed that poverty was caused not by population pressure but by capitalism. However, new-Malthusians such as the **Club of Rome** have updated his principles.

MANGROVE SWAMP HALOSERE

A community of salt-tolerant trees found between 30°N and S of the equator in areas of low wave action where muds can accumulate. The trees have stilt roots to hold the sediments. The system extends up estuaries which are a source of nutrients. The system provides an **ecological niche** for breeding fish. These communities are threatened by tourist developments, marine **pollution** and the loss of the niche. They are in need of **conservation**.

MANN'S MODEL

This model of the structure of a British city was produced in 1965 and was derived from both **Burgess** and **Hoyt**. He saw cities as being the product of both a series of concentric rings of growth and a set of sectors of social and economic activity as shown in Fig. M.1. As a result of his work on several

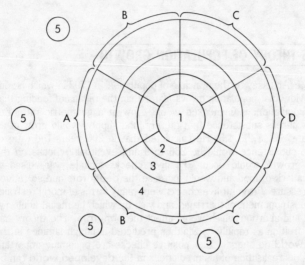

A middle-class sector
B lower-middle-class sector
C working-class sector (and main municipal housing areas)
D industry and lowest working-class areas

1 city centre
2 transitional zone
3 zone of small terraced houses in sectors C and D; larger by-law houses in sector B; large old houses in sector A
4 post −1918 residential areas with post −1945 development mainly on the periphery
5 commuting-distance 'village'

Fig. M.1 Mann's model of the British city

cities he was able to hypothesise that the middle class sector would be on the windward (west) side of the city and the industry on the east together with the lowest social groups. Within each sector he then distinguished the rings in terms of age, style of building and social group. He also noted the presence of **commuter villages** beyond the urban area. Robson produced a similar structure in his studies of Sunderland.

MANTLE

◄ Plate tectonics ►

MARGINAL AREA

Those agricultural areas which operate at the limits of profitability because of the economic costs of production and transport of the product to the existing markets. The margin may also be a physical one where the conditions are at the limits of tolerance of a particular crop or animal.

MARKET AREA

The area served by a particular service which is sometimes known as the **hinterland**. The term is used most to define the area served by a retail outlet.

MARKET AREA ANALYSIS

The type of analysis prepared by retail location consultants to predict the **market area** of a shop on the basis of the cost of the supplies, the prices charged, the distance from consumers and the location of competing shops.

MARKET ECONOMY

An economic system in which buyers and sellers exchange goods for money and in which individuals are free to make decisions about what they will or will not purchase. Similarly firms have freedom to select suppliers. It is the system of capitalism.

MARKET GARDENING

The commercial growing of plants for sale in a small **intensive agricultural** unit. It is called truck farming in the USA. It is more commonly used than **horticulture** and includes open air and **glasshouse crops**.

MARSHLAND HALOSERE

A **succession** found in sheltered areas away from wave action in estuaries, the lee of spits or a bar where accretion takes place. The vegetation promotes the accretion of the tidal muds and is adapted to both salt and tidal inundation. There is a definite zonation from the shore with the oldest areas being well-covered with marsh grass, reeds and sedges and the newest areas merely having algae and halophytes, salt tolerant annuals which trap the silt. The plants grow in the nutrient rich environment. Accretion has been speeded up by the introduction of spartina grass since 1870. The systems are threatened by waste disposal, sewage works, built-up areas, recreational uses of the water and shore and even industrial uses on the shore. Many areas are now **conserved** either because of the landscape value of this wetland ecosystem or because of the migratory wild life which uses the marshes.

MARXIST GEOGRAPHY

The approach to the study of geography which uses the ideas of Karl Marx as the basis of any research. The approach is generally that of **structuralism** in that explanations for the distribution of a phenomenon will be based on the functioning of a market, capitalist system. It is an attempt to understand society as a whole and not just the geographical aspects of society.

MARX'S POLITICAL MODEL OF DEVELOPMENT

Marx said that capitalism would destroy itself because of exploitation, inequality and uncontrolled competition. These forces would lead society towards socialism and, finally, communism when the state would run all aspects of society and its economy for the benefit of all. This progression has often followed a different path because countries have become communist or socialist following revolutions and have skipped the phase of late capitalism, e.g. China. Some states have become socialist by democratic means, e.g. Tanzania and have not progressed through any of the stages suggested by Marx.

MASS MOVEMENT

The downslope movement of **regolith** under the influence of **gravity**. It can be aided by:

internal rock pressure;
opening of joints;
slope angle;
the availability of material to move;
water normally from heavy rainfall;
human actions such as **deforestation**, ploughing before planting, cuttings and quarries;
earthquakes.

See **heave, creep, slide,** flow, **rotational slip, slumping**.

It can be a form of **environmental hazard** because the movement of rock and other material downslope may threaten human settlements. The Aberfan disaster in 1966 was the result of a slide which engulfed a village and its school.

MATERIAL INDEX

An index which shows to what extent the location of a manufacturing industry will be either material orientated or market orientated. It is expressed as the weight of the raw materials divided by the weight of the finished product. If the index is less than 1, the industry will be drawn to the market. In this case the industry is said to be weight gaining and market orientated, e.g. brewing. If the index is greater than 1 and the weight of the raw materials is greatest, it is weight losing and the industry will be located near the raw materials as a material orientated industry.

MAXIMISERS

A general term for the owners or decision makers in economic concerns who act in a rational economic manner to maximise their profits. It is an assumption which is built into the **Von Thünen** model and into models of industrial location such as **Weber** and models of retail shopping behaviour. It is almost impossible for people to act in a totally rational fashion because that presumes that the decision maker has perfect knowledge. People are said, therefore, to be **satisficers**.

MEAN

The sum of values of data in a set divided by the number of values in the data set.

MEANDERS

The sinuous channel plan of a river's bends shown in Fig. M.2. Their

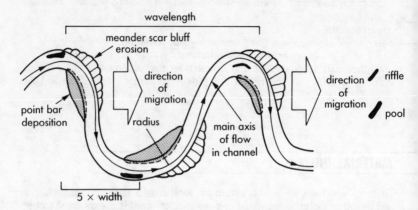

Fig. M.2 Meander terminology

formation is not completely understood but the meander **wavelength** and amplitude are related to a river's discharge. They migrate downstream and increase in amplitude as a result of **bank caving** at the meander scar, point bar deposition and **helicoidal flow**. As the discharge increases so the area of a valley occupied by meanders, the meander belt increases. Meanders can also be incised, cut down into the former valley as a result of rejuvenation. Where meanders have become excessively sinuous they may form cut-offs which striaghten the channel and leave ox bow lakes as a relic of the former meander.

MEAN DEVIATION

Measures the dispersion of a data set, i.e. the average amount that individual values deviate from the **arithmetic mean**. The calculation involves summing the deviations of all values from the mean and dividing by the number of values.

MECHANICAL WEATHERING

The break-up of rocks caused by **exfoliation, freeze-thaw, granular disintegration** and **sheeting**.

MEDIAL MORAINE

◀ Moraine ▶

MEDIAN

The middle value in a data set. It is a measure of central tendency used when data is skewed. Other values on either side of the median can also be identified such as quartiles, a half of the way from the median to the extremes and deciles, a tenth of the way from the median to the extremes.

MEDICAL GEOGRAPHY

The spatial aspects of the study of health and health care. Originally it concentrated upon the distribution of **mortality** from various diseases, a form of study which continues today with the identification of leukaemia deaths around nuclear installations. More recently attention has focussed upon the *delivery* of health care such as innoculation take-up and access to medical facilities.

MEDITERRANEAN CLIMATE

Also called warm temperate western margin climate, it is characterised by hot dry summer months and cool wet winters which are dominated by depressions passing across the area. Only in the Mediterranean countries does the climate extend very far inland and the configuration of the land masses is responsible for this; elsewhere it is very coastal, e.g. California. There are more unique climatic phenomena in the Mediterranean which are caused by the passage of low pressure systems through the area, e.g. the **mistral**.

MEDITERRANEAN SCRUB AND WOODLAND BIOME

A biome associated with the areas of **mediterranean climate** on the western margins of continents. The plant communities are adapted to the climatic pattern of summer drought and winter precipitation. The woodlands vary according to the parent rock material with evergreen oaks and maritime pines dominating. Much of the woodland in the Mediterranean Basin has long been destroyed by people and replaced by cultivated woodland such as the olive tree with a herb layer beneath. Shrub vegetation or maquis is found in the drier, hotter areas where people have interfered with the vegetation on the siliceous soils. Garrigue is the name given to the scrub vegetation found on the limestone soils. In the other areas the vegetation types vary; in California there are species of evergreen oak but the shrub vegetation is known as chaparral. In contrast the West and South Australian areas are characterised by eucalyptus woodland. Most of the vegetation communities in the Mediterranean basin have been affected by human action over the centuries and those areas of 'natural vegetation' such as the garrigue of Provence have probably been influenced by past activities.

MELTWATER

Water originating from the melting of a glacier at its surface, sides or even within the glacier. It can be seen as a **system**. It will form meltwater channels if in its passage away from the ice it has to carve a new valley bacause the old course is still blocked by the ice. Newtondale in the North York Moors is one such channel. The **urstromtäler** of the North German plain are another series of broad valleys formed by streams draining north being diverted by the icefront during the **Quaternary period**. In the latter case the valleys are used by modern communications such as the waterway link between Berlin and the Ruhr.

MENTAL MAP

The **image** of an area and the relationship between elements of that area which is carried in the mind. The map is built up from experience and from other sources such as the media. Inevitably distance is distorted and the retained images are partial and often related to one's own values, i.e. we remember the places which we like and even remember those which we dislike intensely. However, disliked areas are known only as broad areas and without the detail of the favoured areas. Mental maps do often affect other decisions such as where to live and even where to locate a factory.

MERRY-GO-ROUND TRAINS

A form of freight transport for bulk goods over short distances. One example is the journey from a coal mine to a power station when a train working an almost circular route is the cheapest method of supplying the coal.

MERSEYSIDE TASK FORCE

Established following the riots in Toxteth in 1981 this initiative was an example of **targetting**. It was subsumed into other programmes for Liverpool after 1983.

METEOROLOGY

The study of the atmosphere beneath the stratosphere where the surface weather is generated and experienced. It is an increasingly important field of geographical studies given the effects that human activity is having on the operation of atmospheric systems.

METROPOLIS

A term which is reserved for major urban areas which are either the seat of government or a major centre of commercial activity.

METROPOLITAN

Aspects relating to a **metropolis** such as the New York metropolitan region. In some countries the statistical data may be gathered on the basis of Standard Metropolitan Areas and Standard Metropolitan Labour Areas.

METROPOLITAN VILLAGE

◀ Village ▶

MID-OCEANIC RIDGE

Ridges produced in the zone where the convection currents from the earth's mantle rise towards the surface pushing the oceanic plates apart. New crust is created along these ridges and volcanic activity may be associated with them. One such ridge runs north-south approximately in the middle of the Atlantic Ocean and separates the American and African/Eur-Asian Plates. The volcanic activity in Iceland is associated with the **extrusion zone** along this ridge.

MIGRATION

A general term used to describe the movement of population. It is the third significant factor in population change together with **birth rates** and **death rates**. It refers to the permanent or semi-permanent change of residence and should be distinguished from **mobility**. It may be classified according to the type; **international, inter-regional, intra-regional, rural-urban, urban-rural** amd **intra-urban**. It can also be classified by the reasons underlying the decision to migrate; push, pull, political, marriage, labour, persecution and famine or other natural disasters. Some migration is short term or **circulatory** and does not involve the whole family although it involves a stay away from home, e.g. **labour migration**.
◄ Lee's laws of migration, Push-pull model of migration, Stouffer's intervening opportunities ►

MILLION CITY

A city with over a million people. Such cities were concentrated in the mid-latitudes of the northern hemisphere. More recently the new million cities are concentrated in the **developing world**, i.e. more in the tropical regions.

MINIMATA DISEASE

The name of a bay in Japan which has given its name to a disease which killed people living around the bay. The source of the problem was a chemical company discharging mercury chloride into the sea where it was converted into methyl mercury, a highly toxic compound which can enter the foodchain. This it did through the fish and shellfish which were caught by the population thus bringing the accumulated compound to the top of the foodchain. 35,000 people were affected by the disease.

MINORITY

A group of people in a country who are of a different race, language or religion. Minority status is often reinforced by voluntary or enforced **segregation**.

MISTRAL

A cold, dry wind which blows down the Rhône valley from the north when there is a continental air mass lying over central Europe in winter and depressions are passing through the Mediterranean. It may damage vines and fruit if it occurs in the late spring. Shelterbelts are built to protect crops from this wind.

MIXED FARMING

Agriculture which does not depend upon a single crop for the income of the farm. It normally involves both crops and livestock as an interdependent system which is flexible and able to respond more to changes in demand. Much of European farming has depended on a mixture of products from its farms. It is generally regarded as a form of **intensive agriculture**. In the tropics mixed farming also exists and may involve tree crops with a cover crop beneath. Mediterranean agriculture, with its combination of vine olives, intensive vegetable growing and some livestock together with areas of specialised farming such as orange groves, is another form of mixed farming.

MOBILITY

There are several types of mobility; physical mobility or the ability to get about by use of personal or private transport, social mobility or the ability to advance socially without constraint. Labour may be described as mobile if it is prepared to seek out new jobs in fresh locations and capital is mobile if it is able to be moved to investments overseas without exchange controls. Today mobility can also refer to the ability of the telecommunications system to exchange knowledge at speed.

MOBILITY TRANSITION MODEL

A five stage model devised by Zelinsky which relates migration to stages of economic development; each phase has different characteristics:

Phase 1 the pre-modern traditional society with little residential migration and circulation governed by the traditions of the society such as religious visits.

Phase II the early transitional society with large scale rural-urban migration, colonisation of resource frontiers and outflow of **labour migrants**.

Phase III the late transitional society when movement from rural areas and to the resource frontiers begins to decline, emigration declines and more people in the country are mobile.

Phase IV the advanced society with high levels of residential mobility, low rural to urban migration but high urban migration and immigration of unskilled labour migrants from underdeveloped lands.

Phase V a future super-advanced society when migration and circulation are low due to advanced communications, immigration of the unskilled continues and strict control over international migration is introduced.

MODEL

A representation of reality which attempts to summarise the essence of a theory, process, system or an event. It is a simplification of reality which aids understanding and in geography that is the understanding of the complex interactions between people and environments which go to make up the real world.

MODE OF PRODUCTION

A term used in Marxist economics to denote the way in which the productive activities of an economy are organised and how these activities are reflected in social life. Each mode is made up of the means of production and the social relations which derive from that mode of production or the way in which people participate in production. Four modes have been recognised: Primitive Communism, Slavery, Feudalism and Capitalism. There is much discussion on the nature of the last mode because of the variants of it which can be found in the modern world.

MODER

A humus type between mor and mull, i.e. where decomposition is greater than in mor but has not proceeded as far as in mull.

MOLE DRAIN

An agricultural technique used in clay soils to create artificial soil pipes at depth to help in drainage.

MONOCULTURE

The growing of a single crop for sale which invariably leads to disease in the crop, problems with pests and declining fertility if measures are not taken to combat the problems.

MONOPOLISTIC COMPETITION

A market where the sellers have very similar products to sell, e.g. the sale of washing powders. The competition is said to be 'imperfect'.

MONOPOLY

A market which has one seller who controls the provision of a good or goods. The term is used when a firm has an unreasonalby large share of the market.

Most countries have regulations which prevent monopoly, e.g. the Monopolies and Mergers Commission in Great Britain. Monopoly destroys competition which is at the heart of the capitalist system.

MOOR

Open country comprising heathland **ecosystem** found at a higher altitude than **heath**, i.e. over 200 m. The moorland ecosystem is the product of clearance and is often maintained by burning in the interests of the moorland farmers and landowners who use the area for sports such as shooting. Moors may be comprised of heather, bilberry or cotton grass as the dominant species depending on the soil and moisture conditions.

MOR

A type of forest humus in which an abrupt interface with the mineral soil exists showing almost no mixing. Clearly identified plant structures exist at all levels in the decomposition layers.

MORAINES

The general term given to the accumulation of material deposited by a glacier. They can be formed on a glacier, (**supraglacial**), in a glacier, (**englacial**) and beneath the glacier, **subglacial**. **Ground moraine** is the accumulation of drift or till under a glacier which is composed of poorly sorted materials. Linear morainic features include;
Lateral moraines, a ridge of glacial material which lies along the side of a valley glacier and is comprised of debris derived from ice **plucking** of the side and falling debris from the valley side dislodged by **freeze-thaw**;
Medial moraines, a ridge of glacial debris extending down the centre of a glacier caused by the confluence of two lateral moraines when valleys join;
Push moraine, a ridge of glacial debris formed when a glacier readvances across an area of drift pushing it up into ridges as in the Veluwe are of the Netherlands;
Recessional moraine or **stadial moraine**, a moraine developed during a period of standstill in the retreat of a glacier so that a ridge develops across the valley which will vary in height according to the length of the stillstand;
Terminal moraine, the ridge of debris which marks the furthest extent of a glacial advance which may be exceeded by subsequent advances and reworked by subsequent water **erosion**.

MORTALITY RATE

◀ Crude death rate ▶

MOUNTAIN CLIMATE

Climates in mountain areas vary more rapidly vertically than they do latitudinally; within a few thousand metres the climate can change from tropical to the equivalent of the polar areas. Much depends on which mountain area is being studied because all mountain areas show considerable variation in climate between their windward and leeward sides and between valleys. Anabatic and katabatic winds and other phenomena such as the föhn, chinook and Santa Ana have major effects on the weather patterns of mountains and their adjacent lowlands.

MUDFLOW

A process of **mass movement** in which a saturated mass of material moves rapidly, and sometimes catastrophically, downslope coming to rest as a lobe of material which is partially sorted by size of materials.
◀ Mudslide ▶

MUDSLIDE

A process of **mass movement** in which a saturated mass of fine grained silts and clays moves relatively rapidly downslope. It is not as fast a movement as a mudflow. They often occur on motorway cuttings through areas of clay after long periods of rainfall have saturated the soil, e.g. A3(M) at Horndean and the M25 in Kent near Caterham.

MULL

A type of forest humus in which all states of decomposition can be seen from fresh litter at the surface to complete destruction of plant structure at depth. There is high biological activity, an intimate mixture of organic and mineral material, a gradual merging transition with layers below.

MULTILATERAL AID

◀ Aid ▶

MULTI-NATIONAL COMPANY

A firm which manufactures its products or provides its services in several countries but whose control remains firmly within one country. VAG, the Volkswagen Audi Group is a West German multinational manufacturer and Knight, Frank and Rutley are a British service sector multinational. They operate in many countries to avoid tariff barriers and to gain access to cheap labour especially for routine procedure. About 20% of all investment in industry in the world is made by such companies and in more than 20 countries they are responsible for over 33% of the total manufacturing output. Their location requirements are more complex than those of companies operating in one country and often relate to comparative costs and the image of the country as a place of manufacture. They are sometimes called trans-national corporations which is the preferred term of the United Nations.

MULTINATIONAL STATE

A country which contains a range of minority groups such as Yugoslavia or Switzerland. Many such states have problems integrating the various national groups as 1989–90 has shown in Yugoslavia and the USSR.

MULTIPLE NUCLEI MODEL

◄ Harris and Ullman ►

MULTIPLE OCCUPATION

The occupation of a dwelling in an urban area by a number of families. It is normally used to refer to either purpose built flats and apartments or, more critically, to houses and dwellings which have been subdivided to provide dwellings for more households. It is a characteristic of the inner city in many countries in the developed world. It is possible for small factories to be in multiple occupation of a building. Some old mills and factories have been converted to multiple occupancy in order to reuse buildings.

MULTIPLIER EFFECT

The effect that any one economic activity within a region has on other economic activities. The computer assembly worker in Greenock will through his/her spending create other jobs – employment multiplier – either in non-basic activities or in other manufacturing in the region. The new Toyota plant near Derby will need parts which may be supplied from the region and this will have both an employment and an income multiplier effect.

MULTIRACIAL SOCIETY

A society containing a mix of people of different racial origins such as modern Great Britain. Because of the racial undertones in such a term some countries choose to call themselves 'multicultural', e.g. Canada, because this emphasises the more positive aspects of the peoples.

MURDIE'S MODEL

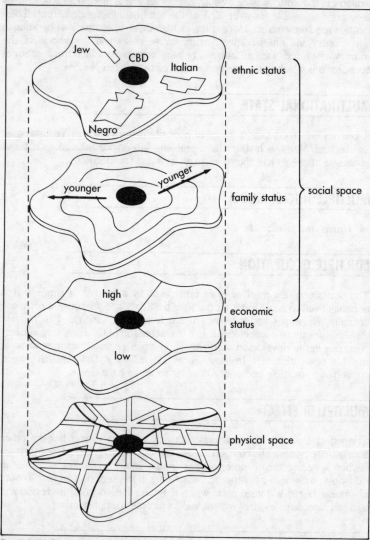

Fig. M.3 Murdie's model of social space: Toronto

Murdie attempted to link the three major elements of social area analysis, i.e. ethnic status, family status, and economic status to physical space based on studies of Toronto. Like other models developed in the mid-twentieth century it arrives at a city structure which combines concentric zones, sectors and distinct districts, see Fig. M.3.

MYRDAL'S MODEL OF CUMULATIVE CAUSATION

The model can be applied at both the national and global scales. The forces of economic growth or decline tend to reinforce themselves so that a new economic activity will create extra jobs both in that activity and in other activities as a result of the multiplier effect. Extra income is created so the chances of further expansion are enhanced and more activities will come to the region. This growth takes place at the expense of other regions because labour migrates to the growing region and more capital is focused upon the growth region. The lagging regions lose labour and capital investment through this backwash effect. The model is circular and cumulative because the pull of capital, resources and labour to the core causes a downward spiral of decline and increasing deprivation in the lagging region. In the global model it is the decline of the less developed countries into increasing poverty and deprivation which is paralleled by a corresponding upward spiral in the developing world.

NATIONALISATION

The state ownership of an economic activity. In many countries utilities such as gas and electricity and rail transport are nationalised. Some countries, notably the UK, have been removing activities from national ownership and selling them to the private sector through the process of **privatisation**.

NATIONAL NATURE RESERVES

Established since 1949 by the Nature Conservancy Council the reserves are designated to facilitate research into and to preserve the natural habitats of the areas. There were 182 reserves in 1982, e.g. Crymlyn Bay, Swansea.

NATIONAL PARKS

Established by the National Parks and Access to Countryside Act 1949, the ten parks cover 9% of England and Wales, see Fig. N.1. This Act has been the major influence on conservation of the landscape in England and Wales. It led to the formation of **national nature reserves**, the body known today as the Countryside Commission, SSSIs and AONBs, as well as the national parks. Each park is administered by a special board and has its own planning powers. Within the parks there is control over most types of development including caravan sites. Afforestation is not controlled and so protest has followed the planting of large areas of the Lake District with conifers rather than deciduous forest. These areas of landscape quality are protected because the landscape would be threatened by development. In the USA, national parks were first designated in 1872 at Yellowstone; the American parks are more definitely wilderness areas than their British counterparts.

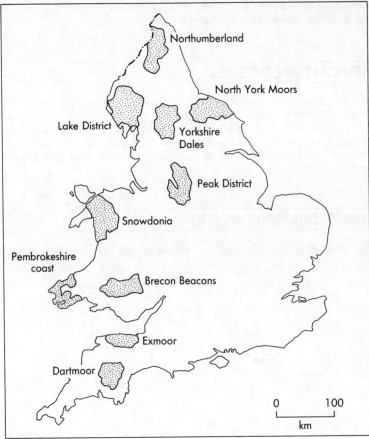

Fig. N.1 National Parks

NATURAL INCREASE

The **crude birth rate** minus the **crude death rate**. It is a measure of population increase. In 1985 the rate of natural increase in Nigeria was 3.1%, Bangladesh 2.8%, Brazil 2.3%, United Kingdom 0.1% and Sweden 0.0%

NATURAL RESOURCES

The parts of the environment which are of use to people, i.e. minerals, energy resources, soils, vegetation such as forests, climate for tourism and even landscapes such as the Lake District are all natural resources. They are subdivided into **renewable** and **non-renewable**.

NEAREST CENTRE HYPOTHESIS

People will always travel to the nearest centre which is offering a good or a service in order to reduce the costs of distance. It is a hypothesis of rational economic persons and has been applied to studies of shopper behaviour. Studies have shown the hypothesis to be false because people are **satisfacers** and they select shopping places on the basis of attributes other than distance from the home.

NEAREST NEIGHBOUR ANALYSIS

The nearest neighbour statistic or analysis ranges from 0.0 = clustered,

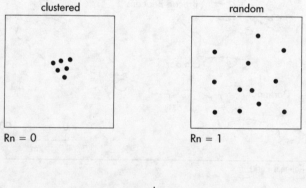

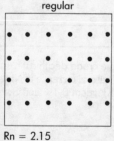

Fig. N.2 Hypothetical settlement patterns. Rn values indicate degree of randomness

through 0.23 = linear clustering, 1.0 = random, to 2.15 = a regular pattern, as shown in Fig. N.2.

- **Stage 1** – Define the area to be studied and locate the pattern of settlements in it. The dots on the edge should either be excluded, in case they are close to others outside the study area, or used only when they are closer to dots in the study area.
- **Stage 2** – Number all the dots and calculate the distance to each nearest neighbour. Add these up to get the observed mean = $\overline{D}o$. The density of points is the number of points ÷ area of the study.
- **Stage 3** – Calculate the expected means for a random distribution:
- **Stage 4** – Now you can calculate:

$$\bar{r}E = \frac{1}{2\sqrt{\text{density}}}$$

- **Stage 5** – Now you can calculate $\dfrac{\overline{D}o}{\bar{r}E}$

Once you have your Rn value you must then ask why the pattern is as it is.

NEIGHBOURHOOD UNIT

A concept used in the planning of **new towns** and other twentieth century suburban areas. It has its origins in the **garden city** and was first used in Radburn, New Jersey. The unit was planned to provide all the essential services needed by a family for their everyday life. The size was normally taken to be large enough to feed a First or Primary School and would contain a range of low order shops providing the everyday necessities of the pre-supermarket and pre-freezer era. In addition each neighbourhood had a church and a public house. The main road system of the settlement formed the boundary between one neighbourhood and the next and this road was often fringed by green zones of parkland. They were units of 2,500 dwellings housing 10,000 people. They were not totally successful because they assumed that everyone would interact within the neighbourhood and not be **satisficers**.

NELLNER'S MODEL

This model is an example of a culturally based model of city structure based on studies undertaken predominantly in West Germany. The model recognises many more zones to the city's development besides the influence of a river on the structure. It also pays more heed to the forces of **decentralisation** with its emphasis on the outer suburban and **commuter village zones,** see Fig. N.3.

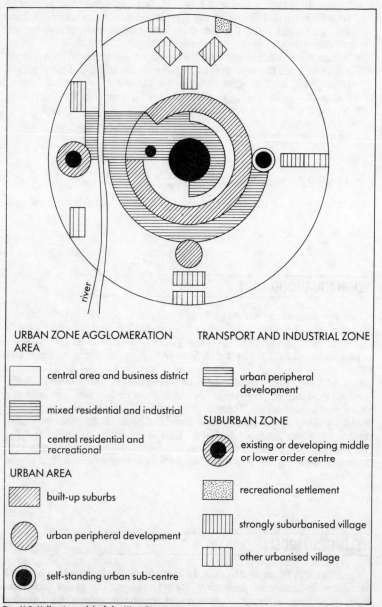

URBAN ZONE AGGLOMERATION AREA

central area and business district

mixed residential and industrial

central residential and recreational

URBAN AREA

built-up suburbs

urban peripheral development

self-standing urban sub-centre

TRANSPORT AND INDUSTRIAL ZONE

urban peripheral development

SUBURBAN ZONE

existing or developing middle or lower order centre

recreational settlement

strongly suburbanised village

other urbanised village

Fig. N.3 Nellner's model of the West European city

NEO-COLONIALISM

The tendency for countries of the **developed world** to maintain their influence over their former colonies. This occurred because the former patterns of trade continued after independence as a result of trade agreements struck at independence. The products of the new states have often continued to be produced by companies whose base is in the developed world, and therefore, the country remains **dependent** in a neo-colonial sense on the former colonial power. Loans made to the developing world originate from the former colonial powers, e.g. France prefers its own former colonies in West Africa for loans and aid.

NEO-NATAL MORTALITY RATE

An **age specific death rate** which measures the deaths during the first four weeks of life per thousand live births. It is a more sensitive measure of health care and the stage of development than the **infant mortality rate** because it is a measure of care at a time when the medical authorities are aware of the risks to the child. Normally reductions in this rate precede those in the mortality rate of the population as a whole.

◄ Neo-natal mortality rate ►

NET PRIMARY PRODUCTIVITY

The rate at which plants are able to store organic matter which is not used by its respiration. It is normally expressed as a weight of dried matter per square unit of area.

NETWORK THEORY

A network is made up of nodes or vertices which are at the meeting point of two or more edges or links. Nodes in transport are railways stations or bus stops whereas the edges are the rail and bus routes. The character of the network depends on the density of the edges and how well the edges are connected to one another, i.e. its **connectivity**. **Accessibility** of each node is measured by the **König number** and the **Shimbel index**.

NEUTRON PROBE

A field instrument which is used to measure the soil water content at a specific depth. An access tube is inserted into the soil down which a neutron source is placed. The amount of neutrons returning to the instrument is measured and by use of a calibration curve moisture content can be calculated.

NEWLY INDUSTRIALISED REGIONS

The term given today to those areas of the country which are being favoured by firms whose products are in the early stages of the **product life cycle**. The regions are also subject to 'property developer hype' so that other jargon terms become commonplace. The M4 Corridor, the Golden Triangle, Silicon Fen, Silicon Glen, and Silicon Valley are some in use in Great Britain. The firms in these regions are often called **sunrise industries**.

NEWLY INDUSTRIALISING COUNTRY (NIC)

Countries whose economic growth rate has been high over the past two decades such as Singapore and South Korea. They have expanded their economic output partly as the result of the **international** spatial **division of labour** and partly as a result of policies of economic growth to encourage indigenous firm formation. The **product life cycle** may be related to the growth of economic activity in the NICs which are labelled LDC in Fig. N.4.

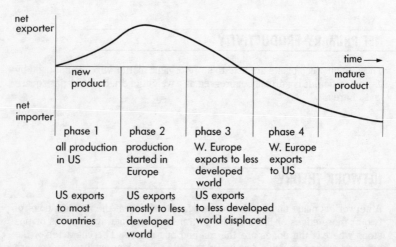

Fig. N.4 The product life-cycle and its suggested effects on US and European production and the less developed world

NEW TOWNS

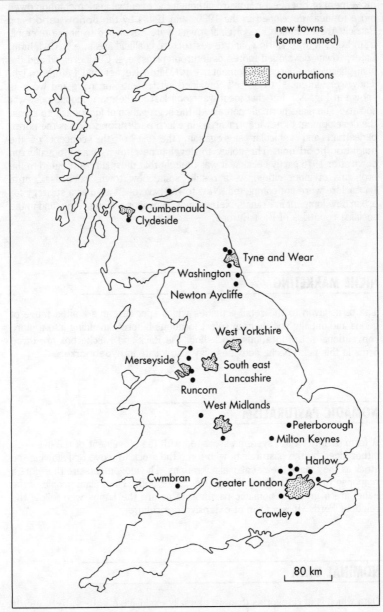

new towns (some named)

conurbations

Cumbernauld
Clydeside

Tyne and Wear
Washington
Newton Aycliffe

West Yorkshire

Merseyside
South east Lancashire

Runcorn

West Midlands

Peterborough
Milton Keynes

Harlow

Cwmbran

Greater London

Crawley

80 km

Fig. N.5

Comprehensively planned new settlements developed after the 1946 New Towns Act. The pressure to build new towns originated from the **garden city movement** of Ebeneezer Howard although the eventual implementation owed more to ideas developed in the 1930s and 1940s by the **Barlow report** and Patrick Abercrombie. The original towns were developed to house overspill from London, e.g. Harlow, and to restructure coalfield areas, e.g. Cwmbran. Later, towns developed more adventurous plans, e.g. Cumbernauld and the final phase were large settlements, e.g. Milton Keynes and Peterborough. They were intended to be self-contained settlements for living and work as shown in Fig. N.5 but that ideal was soon lost in the more mobile post-war society. The plans normally contained: the segregation of functions into areas, the development of new central areas in which **pedestrianisation** is the norm, pedestrian and vehicular segregation, the use of the concept of the **neighbourhood unit**. The policy to develop new towns was gradually run down after 1976 partly because it was thought that they had contributed to the problems of inner cities. As a result, some new towns such as Central Lancashire were not completed. New towns have been used as a strategy for urban development in France, Netherlands, Singapore, Hong Kong, and in the socialist countries of East Europe.

NICHE MARKETING

The term given to the retail activities which specialise in a limited range of goods and initially sell them in select locations before expanding into a more conventional set of locations. Sockshop, Tie Rack and Knickerbox are three firms in this field. Niche activities may also locate in **hypermarkets**.

NOMADIC PASTURALISM

A form of agricultural system associated with the movement of the livestock either aided by the pasturalists who force the stock to move or following the stock as they make their seasonal migrations. The movements are therefore a response to either food supplies or climate or both. The Fulani people of the Sahel are a group of nomadic pasturalists as are the Lapps who follow the reindeer herds. It is a form of **extensive agriculture**.

NOMINAL DATA

Data which is in categories that are mutually exclusive e.g. 0–4, 5–9, etc. It is also known as categorical data.

NON-BASIC ACTIVITY

Those activities whose product is used within an area and which are dependent on the area's internal market. It is often subdivided into producer and consumer serving activities. It is also called **residentiary activity**.

NON-PARAMETRIC STATISTICS

The tests which are used on data obtained with **nominal data,** i.e. in categories which are mutually exclusive or with **ordinal** scales of measurement, i.e. where data is ranked. They are also used if there is uncertainty whether the population is normally distributed because there are abnormally high values. Some non-parametric tests are Mann-Whitney U test, **Spearman rank** and Chi_2.

NON-RECYCLABLE RESOURCES

◀ Non-renewable resources ▶

NON-RENEWABLE RESOURCES

Resources which have been built up over geological time and where any use will deplete the stocks because of the length of time required to renew the source. All fossil fuels and minerals are non-renewable because the resource is finite and can only be used so long as there is a resource to exploit. Non-renewable resources may be subdivided into **non-recyclable,** e.g. fossil fuels which cannot be renewed, and **recyclable** which may be reused and recycled, e.g. lead and copper.

NON-STOCKABLE RESOURCES

◀ Renewable resources ▶

NORMAL DISTRIBUTION

A frequency distribution which when plotted on graph paper would show a bell shape with the majority of values lying close to the mean. Parametric tests assume such a distribution.

NORTH AMERICAN CITY

◀ Wreford-Watson's model ▶

NUCLEAR POWER

◀ Atomic power, Energy ▶

NUTRIENT CYCLING

The means by which the chemicals essential for the existence of organisms are moved through the **ecosystem** from their presence in the **primary producers**, through the **trophic levels** returning to the soil by **decay**. Once returned to the soil the nutrients may be recycled. Agriculture may destroy the cycle because it takes nutrients away in the form of crops. But it is possible to restore the nutrients either by careful husbandry involving fallow periods or by the use of artificial or natural fertilisers. The main nutrients which are cycled are carbon, phosphorous, calcium, sulphur and water.

NUTRIENT STORE

Associated with the **cation exchange capacity**, the net negative charge of soil material provides sites on which nutrients (cations of K, Mg, Ca etc.) can be held or stored and hence be available for plant growth. Many of these nutrients taken up by the plants are eventually returned to the soil when the plants dies and decomposes.

OASIS

A place in the desert which has water which enables plant growth to take place. **Deflation** has created hollows which reach down towards the water table so that shallow wells give access to the water. In other cases the water is supplied by **Artesian Wells**.

OCEAN CURRENTS

The movement of the water in the oceans in a pattern of general circulation caused by the effects of the earth's rotation, the effects of the general circulation of winds and the pattern of land masses. There are *zones of upwelling* within the oceans where waters rise from depths and *zones of convergence* where waters descend and move parallel to one another, often separated by a countercurrent.

OCEANIC TRENCH

◄ Subduction zone ►

OCEANOGRAPHY

The study of oceans and seas including the natural life in the oceans. In geography it has normally been confined to the study of ocean currents and waves and the oceans as a resource.

OFFICE DECENTRALISATION

The process of moving offices from major metropolitan city centres to other locations in the suburbs or to other towns and cities in the national urban system. In Great Britain this process was officially encouraged by a government body, the **Location of Offices Bureau (LOB)** between 1964 and 1979. Decentralisation still continues due to pressure from rents, space needs, labour costs, and takeovers and mergers. Traffic congestion is a further pressure.

OFFICE DEVELOPMENT PERMIT (ODP)

A scheme, similar in its operation to the **industrial development certificate**, which was designed to steer office employment to the regions. The area of control in which permits were required varied over the years 1965–79 as did the size of building which required a permit.

OFFICES

The buildings in which certain of the tertiary activities associated with **producer services** take place. They may contain organisations ranging from international headquarters of a firm to the office of an accountant in a small town. Their location has been studied to show how there are forces which have encouraged their development in the **CBD** and other nodes within an urban area. **Agglomeration** of offices occurs because of particular linkages which can be complementary, e.g. the use of common advertising facilities; ancillary, e.g. the provision of joint lunch facilities; supply, e.g. the use of common office suppliers; and information links, e.g. the acquisition of common knowledge without payment.

OFFSHORE GRAVEL

Gravel which is dredged from the sea bed for use in the construction industry. Its source is either **drowned beaches** or, more probably, submerged river terraces. The former terraces of streams which drained out from Hampshire into the English Channel are the source of much offshore gravel brought ashore along the Sussex and Hampshire coasts. Some people are of the opinion that this removal of sands and gravels is altering the supply of material to the **beach store** and leading to more gravel rather than sandy beaches on the south coast.

OLIGOPOLY

An economy where a small numer of firms control much of the output and employment. Chemical production in West Germany is dominated by three oligopolistic companies.

OPEC

This is an acronym for the Organisation of Petroleum Exporting Countries.
◄ Cartel ►

OPEN-CAST MINING

The excavation of raw materials by taking away the overlying surface soil and rocks rather than mining the materials from shafts sunk into the ground. Brown coal is mined in pits the size of a major city and up to 600 metres deep in West Germany.

OPTIMISERS

The people who make optimal decisions based on near perfect knowledge. It is similar to the idea of rational economic people and has been discredited by behavioural research.

OPTIMUM POPULATION

A population existing in an area where the output per capita is maximised within the prevailing technological and socio-economic constraints of the area as shown in Fig. O.1. The optimum is below the survival level where the

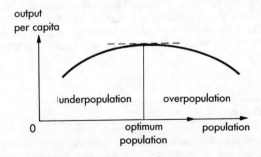

Fig. O.1 The relationship between population and output

resources are unable to support the population and **overpopulation** results. **Underpopulation** is any state below the optimum. All of these are difficult to measure in absolute terms because of the changing resource base and the difficulty of measuring the resource value of certain economic activities such as services.

ORDINAL DATA

Data where the observations are placed on a scale or ranked. Each observation is unique in terms of its position on the scale. It is sometimes called a ranking scale. Ordinal data is frequently used in perception studies.

ORGANIC WEATHERING

The breakdown of rocks by plants and animals. Tree roots and burrowing animals can break up rocks. The **decomposition** of plants creates organic acids which chemically decompose the rocks. Lichens on the stonework of old buildings also have a similar effect.

OROGENY

A major period of mountain building. Three major orogenies have affected Europe; the Caledonian, the Hercynian and the Alpine.

OUTPORT

A port facility which has been developed down-river from the original port because ships have increased in size, estuaries have become silted up and large areas of land are needed for onshore storage of, e.g. containers. It is a part of the general downstream movement of port facilities outlined in **Bird's** Anyport model.

OUTWASH PLAIN

An area of **fluvio-glacial** material which has been built from material carried away from the glaciers by **meltwater**. It will contain a series of fluvio-glacial landforms; **kame, esker, kettle holes**, and reworked **moraines**.

OVER–GRAZING

Where too many animals are grazing an area of land and the vegetation is not replaced as fast as it is consumed. The result is the removal of the vegetation which enables the agents of **soil erosion** to remove the soil. It is of particular danger in areas of seasonal rainfall where the vegetation is eaten away and trampled around water-holes.

OVERLAND FLOW

A part of the **hydrological cycle** and the **catchment basin system** which refers to the movement of water across the surface of the earth. The water has fallen as precipitation and has not been **intercepted** although some will have reached the surface as a result of **stemflow** and **drip**. The greater the intensity of precipitation the greater the overland flow. It also increases once the **infiltration rate** of soil has been exceeded. Overland flow increases down a slope and is responsible for the movement of soil particles especially on bare slopes leading to **soil erosion**. **Throughflow** can return to the surface as a result of the build up of a saturated wedge of soil water at the base of a slope.

OVERPOPULATION

Occurs when the resources are unable to sustain a population at its existing living standard. It is a state which is above the **optimum population** and below the **survival level**.

OVERPRODUCTION

Food production in an area which is in excess of that country's requirements. It occurs because, countries are anxious to obtain export earnings, subsidies encourage overproduction, the physical conditions are ideal for a crop, the technologies have improved output and more land can be used because of irrigation. Overproduction may be curbed by; policies such as **set- aside**, halting grants for increasing the agricultural area, be selling surpluses cheaply to **aid** developing countries, by subsidised selling as a part of a political policy e.g. to East Europe. Overproduction is a grave threat to the environment because it can lead to **soil erosion**, **pollution** e.g. from nitrogenous fertilisers and to the excessive use of **energy**, water and soil **resources**.

OXIDATION

A form of **chemical weathering** in which the oxygen in water reacts with rocks containing iron to form brownish oxides and hydroxides.

OXISOL

A term used in the US soil taxonomy for soils containing an oxic horizon within 2m of the surface. An oxic horizon is a layer characterised by the lack of 2:1 lattice clays with only 1:1 kaolinite clays present, low CEC, high presence of iron, aluminium, hydrated oxides and insoluble quartz.

OZONE LAYER

A layer where ozone is concentrated at a ratio of ten parts per million in the stratosphere. Normally the layer lies between 25 and 35 km above the earth's surface. It absorbs incoming ultra-violet radiation which leads to the warming of the stratosphere.

The ozone layer is disturbed by chlorofluorocarbon gases (CFCs) because ultra-violet light causes the chlorine to destroy ozone molecules (1 chlorine molecule can destroy up to 100,000 ozone molecules). CFCs are used in aerosols – 62 per cent of the 800 million produced in the UK in 1988 used them – in plastic foams, refrigerants, and in fire extinguishers (similar gases called halons).

▶ EVIDENCE

- the destruction is most 'visible' over Antarctica when 50% of the layer was destroyed in 1987
- NASA notes a decline of 3% per annum in ozone over the northern hemisphere, which is four times faster than previous predictions
- slight evidence for a 'hole' over Spitzbergen

▶ EFFECTS

- ultra-violet radiation increasing – destroys marine and terrestrial food chains – could affect fishing and agriculture
- increase in skin cancers due to higher ultra-violet doses
- implicated in the **greenhouse effect** because CFCs also limit the heat leaving the atmosphere

▶ REACTIONS

- the UN Environment Programme (1980) first called for reduction; the Montreal Protocol (1987) called for a 35% cut by 1999
- toleration of continued use of CFCs in **developing world** until they can afford to change: this is a problem, because greatest potential growth is here
- increased control over use is a reaction to latest evidence – 1989 London Conference and Hague Conference – speeding-up changes

PAMPAS

◀ Temperate grassland biome ▶

PARADIGM

The assumptions, methods of investigation and conclusions which are commonly accepted by a group of geographers (or other scientists) during a particular period and so define the dominant forms of study. These changes over time and shifts of paradigm often represent major advances in the study of a discipline. The **quantitative revolution** is the term given to one shift which occurred in the 1960s when there was a move from a regionally dominated, descriptive discipline to a study of geography based on the **scientific method** and **spatial analysis**. It is possible for several paradigms to be in use at the same time in a discipline and this is the case in geography where research is not dominated by any one paradigm.

PARALLEL SLOPE RETREAT

◀ Penck's slope replacement model ▶

PARTICLE SIZE

This is the size of the individual inorganic particles normally defined the particle's 'b' axis measurement. The term *particle size distribution* is normally used to describe sediments, however it can be used in soils and in this case the distribution defines the soil texture.

PARTNERSHIP AUTHORITIES

Set up by the Inner Urban Areas Act 1978 to provide high levels of support for new developments to improve the economy of areas such as Hackney. The seven partnerships were between central government (i.e. the Departments of the Environment, Trade & Industry) and local government. **Programme authorities** and **designated districts** were part of the same Act.

PATTERNED GROUND

A broad term which refers to the approximately symmetrical forms which are found mainly in areas subject to frost action. It is a form of **congeliturbation** which results in **stone stripes**, stones aligned in paralled lines down a slope; **stone circles**, stones sorted into circles; **stone nets**, stones in a circular or polygonal mesh on almost level ground which are often called **stone polygons**.

PEAT

An unconsolidated material made up almost entirely of undecomposed, or partially decomposed, organic matter accumulating under conditions of excessive moisture.

PED

A naturally occurring unit of soil structure such as an aggregate, crumb, prism, etc. It is *not* a clod of soil, which is formed artificially, normally by mechanical means, e.g. ploughing.

PEDALFER

A now obsolete term in which the oxides of iron and aluminium (sesquioxides) increase relative to silica during soil formation.

PEDESTRIANISATION

The complete exclusion of traffic from streets normally for the benefit of a retail area or a tourist district. City centre retail districts were first pedestrianised in the **new towns** although many other towns and cities followed suit often in response to threats to the trading pre-eminence of the high street from out-of-town retailing. Some cities permit public transport into the pedestrian area and some others, particularly in West Germany permit delivery and service vehicles into the area until, e.g. 10 a.m. Pedestrianisation often requires expensive access to be built to the rear of properties involving **comprehensive redevelopment**. Today pedestrianisation is being used to make surburban housing districts more liveable by making vehicular access and, particularly, short cuts through an area, more difficult. The Germans call this 'traffic calming'.

PEDIMENT

The low angle slope which is formed as a result of **paralled slope retreat** at the base of the **cliff** and **debris slopes**. At the upper edge there is a marked increase in slope angle to the cliff and debris slope. It is found at the base of mesas and buttes.

PEDIPLAIN

A low angle surface which was probably formed through the coalescence of two or more **pediments**. It is the final stage of the process of **parallel slope retreat**.

PEDOCAL

A now obsolete term for soils in which calcium accumulates during formation.

PEDOGENESIS

Soil genesis, i.e. the processes associated with the evolution of the soil profile.

PENCK'S SLOPE REPLACEMENT MODEL

A sequence of slope replacement which results in gentler slopes with lower angles. A **cliff slope** will be replaced by a **debris slope** which is in turn replaced with a **foot slope**.

PEOPLE-MADE SOIL

Called also anthroprogenic soil or *plaggen* soil (a dutch term). Formed through continuous additions of organic and inorganic material to the surface of the soil until the depth is so great that the properties of the soil are independent of the original parent material.

PERCEIVED CAPACITY

The individual consumer's own view of the capacity of a tourist or recreation resource beyond which it is deemed to have a negative utility. Once this stage is reached the resource no longer provides enjoyment.

PERCEPTION

◀ Behavioural geography ▶

PERCOLATION

The vertical movement of water through the soil following **infiltration**. It is the stage in the **hydrological cycle** when the water moves from the soil down through the lower **regolith** and into the **groundwater store**.

PERIGLACIATION

The processes which occur in those areas adjacent to glaciers and in those areas where there are cold climates but no glaciers. They overlap with **permafrost** environments. It is a rather vague term referring to an indeterminate area. Evidence of periglaciation can be found in relic form in, e.g. **dry valleys** in the South Downs.

PERI-NATAL MORTALITY

Deaths occuring either to an unborn baby (foetus) or during labour and birth. It can be used as a measure of health care and the economic and social development of a population.

PERIPHERALITY

The feeling of remoteness which is experienced by areas which are not located in a core economic region either in a country or a trade bloc. It is one of the major problems which faces southern Italy and Western Ireland within the **European Community**. The South West and West Wales are two British examples. Northern Ireland is peripheral within the European Community and Great Britain but within the province, the Londonderry area is regarded as peripheral and Belfast as central. Therefore the term is a relative one.
◀ Friedmann's centre-periphery model ▶

PERMAFROST ENVIRONMENT

An area in which the temperature of the zone beneath the surface has remained at or below zero for at least two years. It is a zone of permenantly frozen ground beneath an active layer in which **frost cracking, frost heave, frost wedging,** and **freeze-thaw** take place.
◀ Lenses, Patterned ground, Stone circles, Stone polygons, Stone stripes, Wedges ▶

pH METER

Instrument used to measure soil acidity. It is also possible to measure soil pH by colour tests.

PHOTOSYNTHESIS

The process by which the cells of green plants utilise sunlight together with carbon dioxide and water to produce oxygen and the food molecules it needs. It occurs in the **primary producers** and results in the production of plant materials or **biomass**.

PHYSICAL CAPACITY

The ability of a **tourist** and **recreation resource** to accomodate visitors. The capacity will vary with the activity and once the capacity is reached then the visit becomes less pleasurable. Physical capacity can be that of a car park, the number in a stadium or the number who can pass through HMS Victory in a given period.

PHYSICAL GEOGRAPHY

The study of the processes which create, modify and destroy the natural environment. It is traditionally divided into three; the **atmosphere**, the **biosphere** and **lithosphere** and the various branches of physical geography study elements of the threefold division. The branches are; **geomorphology**, hydrology, oceanography, meteorology, climatology, pedology and **biogeography**. Today most geographers accept that the processes operating in the natural environment are influenced by human actions either directly, as in the case of **deforestation**, or indirectly as in the case of **slumping** in road cuttings. Modern geography attempts to study the **interface** between the natural and human environments and so physical geography, unless it takes into account the influence of the environment on human actions and the influence of people on environment, is often regarded as an outmoded discipline.

PICTOGRAMS

Maps and diagrams which make use of symbols that show the nature of the variable being measured. They are a somewhat journalistic form of data portrayal because they are hard to measure accurately although they do give a general impression of the data.

PIE GRAPHS

A circular area which is divided into sections as in Fig. P.1, showing the proportions of a value which are made up of component categories. The circles may have varying sizes and be located just like a **proportional located symbol map**. The size of a component category is calculated by dividing the observed category size by the total sample size and multiplying the result by 360; this gives the number of degrees of the angle of the segment to be inserted in the circle. Pie graphs are a simple method of portraying data and at A and AS levels you should be expecting to use more sophisticated methods wherever possible.

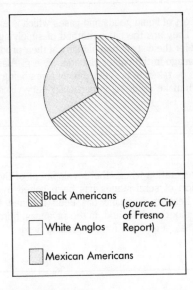

Black Americans

White Anglos

Mexican Americans

(*source*: City of Fresno Report)

Fig. P.1 Population composition in Fresno, USA

PIGGY BACK SYSTEMS

The broad term given to various methods by which the trailers of one means of transport are carried by another. Motor rail is the most obvious case used by people as was the rail ferry in the past. Today, road transporters are frequently carried overnight by rail across West Germany and there is also a scheme to load barges or lighters aboard ships known by its acronym LASH. RoRo (Roll on Roll off) is another system. The advantages of both modes of transport are maintained over greater distances, eliminating the delays at break of bulk or transhipment points.

PINGOS

Large hillocks found in **periglacial** regions which may rise to 65 m. They are formed either as the result of the growth of a frozen mass of water in a former lake which has been filled with silt or by the movement of groundwater up into a frozen sub-soil as a result of **artesian** pressure which then causes the surface to balloon upwards into a dome or hillock.

PIPELINE

A method of transport of liquid goods and gases which is expensive to lay but cheap to operate. They are the basic method of supplying oil refineries and chemical plants and for the distribution of many of their products. Pipelines are more difficult to manage in the colder climates of, e.g. Alaska, where special precautions have to be taken to prevent the contents being slowed by the low temperatures which make crude oil in particular very viscous.

PLAGIOCLIMAX

A form of climax vegetation where the development of the plant community through a **succession** of seral stages has been halted by human action. In Great Britain the heather moorland has been retained by burning in the interests of grouse moor owners, and, in the savannas, burning is also used to maintain such a climax vegetation for the grazing herds.

PLANNING

The process of achieving a particular goal on the basis of a set of actions designed to achieve that goal. There are various scales of planning ranging from national to local. Some forms of planning do not have a spatial or geographical aspect although they can result in spatial effects, e.g. social security plans which result in homelessness. National planning has been favoured in socialist states although other states such as France have had national plans since 1945. The national plans are normally related to **regional plans** although in the case of Britain regional planning from 1964 until 1979 was only rarely related to an overall national strategy. Since 1979 the work of the Regional Economic Planning Councils has been abolished and replaced by less effective advisory bodies, e.g. SERPLAN (South East Region Planning group). Interregional planning in Britain has been left to policies such as those for industrial and office location and the **new towns**. Planning may also be classified as i) economic, ii) social and iii) physical. Economic planning is normally administered at the regional scale while social and physical planning are intertwined in the local planning process of **town and country** planning.

◀ District plan, Regional planning, Structure plan, Town and country ▶

PLANNING GAIN

The outcome of bargaining in the planning process whereby the developer of a site will offer the local authority some benefits in order to gain planning permission. Many retail developments have provided improved transport access in the vicinity of the store, e.g. car parking not only at the store but also in adjacent city or town car parks.

PLANTATION AGRICULTURE

A commercial farming system associated mainly with the colonial period in the tropics. It involves the growing of a single crop for profit, with high initial capital input, low cost intensive labour input and a highly profitable output of specialist and, sometimes, luxury crops, e.g. Rubber in Malaysia and tea in Ceylon. After independence many areas growing such crops developed a **neo-colonial** relationship because of their continued dependence on the crop. However, many countries diversified the crops and reduced the importance of the plantation system for the economy.

PLATE TECTONICS

The explanation of the structure of the earth's crust which suggests the mechanisms by which the plates were formed and have moved. There are 7 major and 12 minor plates as shown in Fig. P.2 which are moved by convection currents within the **mantle**. The plates are either continental and formed of sial granitic material rich in silica and alumina which is lighter than the second type, the oceanic which are formed of the heavier sima, rocks rich in silica and magnesium. Some plates are a mixture of both types.

The edges of the plates are known as plate margins. These can be classified into **constructive margins** where the ocean crust plate is being built up on each side of an oceanic ridge; **destructive margins** where the ocean floor is being **subducted** and lost beneath a continental plate and **conservative margins** where plates are sliding past one another along **transform faults**. Destructive margins may occur between two oceanic plates and lead to the formation of an **island arc** and between two continental plates leading to the formation of mountain ranges such as the Himalayas. Plate tectonics helps to explain the pattern of **earthquakes**, volcanic eruptions, and **volcanoes** past and present. It is also the basis of the theory of **continental drift**.

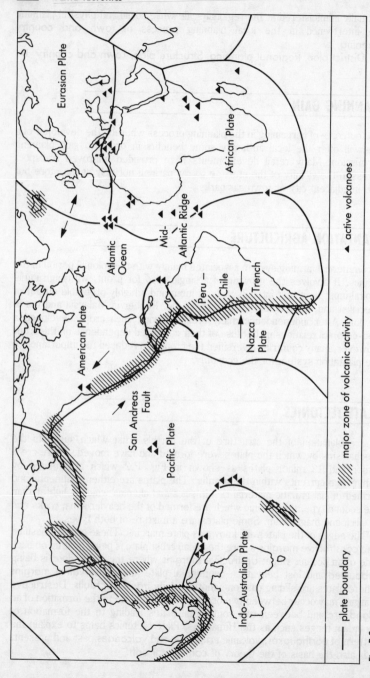

Eurasian Plate

African Plate

Atlantic Ocean

Mid-Atlantic Ridge

Peru – Chile Trench

Nazca Plate

American Plate

San Andreas Fault

Pacific Plate

Indo-Australian Plate

▲ active volcanoes

////// major zone of volcanic activity

— plate boundary

Fig. P.2

PLINIAN ERUPTION

◀ Volcano ▶

PLOUGH PAN

A sub-surface layer of the soil with dense massive structure and hence low permeability formed by the induced pressure at the base of the plough. The plough pan therefore defines the depth of ploughing.

PLUCKING

A process of glacial **erosion** in which blocks are removed from the rock floor and sides by ice which has frozen to them. As the glacier moves so the rock is pulled away and transported downstream, being used as an agent of abrasion.

◀ Cirque ▶

PLUTON

◀ Batholith ▶

POACHING

The compaction of the soil caused by the use of heavy machinery and animal hooves which leads to problems in **infiltration**.

PODSOLIC SOILS

Soils which are acid versions of the **brown earths** tending to form on acid parent materials. Some evidence of Fe and Al movement may be seen, i.e. a few bleach quartz grains in the top horizons. However, there is not enough accumulation of Fe and Al at depth to make them true podsols.

PODSOLISATION

The process that leads to the formation of a podsol soil type. Most soil classification schemes define a podsol as a soil with a horizon at depth enriched with iron and aluminium oxides. Such a horizon is known as a spodic horizon (hence, *spodosols*, a US term for podsols). Podsolisation thus involves the mobilization of the Fe and Al in the upper part of soil and its subsequent immobilisation and accumulation at depth in the spodic B (B_s). All podsols show this B_s horizon, however, many variations of podsols exist reflecting the variable intensity of the process. At its most extreme all Fe and Al is removed from the top part of the B leaving only a layer of bleach quartz sand (albic horizon). Below this a thin black humic layer exists of mobilised organic compounds, followed by a hard indurated iron pan before the spodic B_s is reached. See Fig. G.4 on page 103.

POINT BAR DEPOSITION

◀ Meanders ▶

POLAR CONTINENTAL AIR MASS

In the UK case this is air which originates over Eurasia during the winter months which brings cold conditions particularly to the east where snow showers will result. In summer the air mass is warm and dry although the moisture which it gathers while crossing the seas makes it less stable due to the cooling effect of the seas on the lower layers. Inland areas tend to be clear and warm.

POLARISATION

◀ Growth pole theory ▶

POLAR MARITIME AIR MASS

The most common air to affect the UK which has its origins over the North Atlantic and Greenland. It is unstable cold/cool air which has been heated in its lower layers by the warmer ocean giving rise to showers. Very often the Polar maritime air has been drawn around a depression and approaches the UK from the south-west. In these cases it is highly modified and normally called returning polar maritime; it is much warmer and more unstable.

POLITICAL CAPACITY

The acceptable capacity of a tourist or recreation resource which provides the maximum benefit to all visitors. It must be decided by management because each group has its own **perceived capacity**; the elderly might perceive capacity to have been reached before the parents of a small family.

POLITICAL GEOGRAPHY

The study of political phenomena such as the state and local government, and the ways in which they affect decision making. The role of government in many aspects of human life ranging from regional planning to planning zonation and the allocation of council housing ensures that the political aspects of the making and changing of landscapes cannot be ignored in modern geography.

POLJE

A large surface depression found in areas of **karst** whose form is linear with its longer axis being aligned with the grain of **carboniferous limestone**. They were formed as a combined result of **solution**, faulting and water **erosion** when the water table was higher. All of these processes lead to the collapse of an **underground cave system** to form a polje.

POLLUTION

The deliberate or accidental introduction into the environment of a gas, fluid or solid substance which will endanger health and harm the **ecosystem** of the area. It will also reduce the amenity of the area by its presence. There are three main types;
- *air pollution* such as the emission of lead from car engines, carbon dioxides from cars, sulphur from combustion, smog and aerosols;
- *water pollution* from sewage effluent, slurry, and fertilisers draining into streams, *thermal pollution* and oceanic pollution from oil **spills** and waste disposal;
- *land pollution* especially as the result of the legal and illegal dumping of toxic waste.

Some would add a fourth category, that of *noise pollution* from proximity to road traffic, a form of pollution which has resulted in the residents along the new A27 between Havant and Chichester protesting to the Minister of the Environment.

POOL

An erosional feature which is associated with **riffles** and thought to be associated with the development of **meanders**.

POPULATION DENSITY

The number of people per unit of area, normally per square kilometre or per hectare. Density is influenced by a number of physical and human factors which vary between the **developed** and **developing worlds**.

POPULATION DEVELOPMENT MODEL

◀ Demographic transition model ▶

POPULATION EXPLOSION

A popular term for the rapid or exponential growth of population caused by a sudden decrease in the **death rate** and the simultaneous increase in the **birth rate**. It is associated with Stage 2, **early expanding stage**, of the **demographic transition model**. A population explosion is a cause for concern when there are not the food resources to sustain the increased population and it is that concern which was the focus of **Malthus** and, more recently, the **Club of Rome**.

POPULATION GEOGRAPHY

The study of the growth, distribution and movement of population. It was a study which concerned itself with distribution but increasingly, along with other social sciences, geographers have been examining patterns of fertility, **mortality** and **migration** at a variety of scales. This has led some into the study of models of population growth and change, and patterns of disease mortality and health care.

POPULATION POLICY

The measures taken by a government to cope with the issues raised by changing population numbers and resource base in a country. The measures are designed to affect population change. Policies are either **expansionist;** restricting the availability of birth control, outlawing abortion, granting family allowances to encourage childbirth, supporting working mothers, improving health care and restricting the roles of woman in society, or **control policies;** introducing birth control and permitting abortion, encouraging sterilisation, introducing controls over the number of children per family, reducing support for large families and by campaigns to persuade people to limit family size.

POPULATION PRESSURE

Population growth which is in excess of the capability of the economic and social system of a country. When the pressure is too great the **survival level** has been reached.

POPULATION PYRAMID

◀ Age-sex pyramid ▶

POPULATION–RESOURCE RATIO

The relationship between the amount of and quality of a country's or an area's **natural resources** and the size and competence of its population. The ratio enabled Ackerman to distinguish five types of **population resource region**:
- the **United states type**, advanced, technological societies with low population growth and high availability of resources;
- the **European type**, advanced technological societies with a high ratio;
- the **Brazil type**, technologically deficient regions of low population resource ratio;
- the **Egyptian type**, regions of low population resource ratios; and
- the **Arctic-desert type**, areas of poor technology and limited food-producing resources.

The classification is dated due to new discoveries of resources, e.g. Alaskan oil.

POPULATION RESOURCE REGIONS

◀ Population–resource ratio ▶

POSITIVE CHECKS

◀ Malthus's theory of population growth ▶

POTENTIAL ENERGY

◀ Energy ▶

PRAIRIES

◀ Temperate grassland biome ▶

PRECONDITIONS FOR TAKE-OFF

◀ Rostow's stages of economic growth model ▶

PRED'S BEHAVIOURAL MATRIX 1967

A means of analysing locational decision making which is represented in the form of a matrix. It assumes that over time the decision makers are able to gain more information and better information on a set of locations and that their ability to use the information which they have received will improve.

PRE-MODERN TRADITIONAL SOCIETY

◀ Mobility transition model ▶

PRESSURE GROUP

A voluntary body of people who come together to influence decision makers at all levels of government to act in a particular manner which serves the group's interest best. Many groups begin as interest groups but become pressure groups when that interest is threatened. Some will form to oppose a new road while others such as the Council for the Preservation of Rural England are permanent. There are even semi-governmental pressure groups such as the Nature Conservancy Council.

PREVENTATIVE CHECKS

◀ Malthus's theory of population growth ▶

PRIMARY DATA

Data collected from original sources such as measurements of a stream's discharge or land use in a city centre. It is data gathered in the field.

PRIMARY ENERGY

The original sources of energy before conversion for human use. They are measured either in terms of their calorific value or in tonnes of coal equivalent. Coal, natural gas, oil, the sun, water, timber and peat are all primary energy sources.

PRIMARY PRODUCER

The first trophic level in an ecosystem where energy is fixed by photosynthesis. The plants are autotrophs which use the carbon dioxide together with the sun's phototrophic energy and inorganic elements (chemotrophic) for energy.

PRIMARY SECTOR

That sector of the economy involved with the production, collection and use of natural resources. It includes agriculture, fishing, forestry and mining.

PRIMARY TOURIST RESOURCE

The basic resource on which tourism depends such as coastline, the countryside or the mountains.

PRIMATE CITY

This is city which dominates the rank–size relationship to a disproportionate extent, e.g. Lima within Peru. The term was first used by Jefferson in 1939 to note cities which dominate the economic, political and social life of countries. They are normally associated with small countries or islands, e.g. St Peter Port in Guernsey, the export orientation of the country, e.g. Maracaibo, Venezuela and the colonial past of a country, e.g. Kampala, Uganda.

PRISERE

The initial stage in the plant succession leading to the development of a sere.

PRIVATE VENTURE CAPITAL ORGANISATIONS

New private sector initiatives developed in the late 1980s to aid investment in particular areas, e.g Northern Investors, Newcastle upon Tyne.

PRIVATISATION

A policy which is designed to take economic enterprise and social services away from the public sector and place them in private ownership. It is a policy which has been followed in Great Britain since 1979 and has resulted in the selling to the private sector of firms in steel making, car assembly and shipbuilding as well as utilities such as gas and electricity. Even water supply and waste disposal have been sold. The motive is partly an ideological one but partly to provide alternative funds to taxation so that taxation may be reduced.

PROCESS LINKAGES

These are the links between firms and their suppliers, (backward linkage) and their consumer companies (forward linkages).

PROCUREMENT POLICY

The policy of the government towards its own purchases of goods and services which can influence the distribution of economic activity. In particular, defence expenditure can help regions, e.g. naval vessel construction on Tyneside or at Barrow in Furness. However it has been shown that modern defence expenditure is focused very strongly upon the South East and South West and is of a greater value than all regional aid given to the Northern regions, Wales and Scotland.

PRODUCER SERVICES

A subdivision of the tertiary and quaternary sectors of employment which involves the servicing of manufacturing. It includes accountants, legal services, banking and finance and business analysts.

PRODUCT LIFE CYCLE MODEL

The five stages, (see Fig. P.3) in the life cycle of a manufactured product:

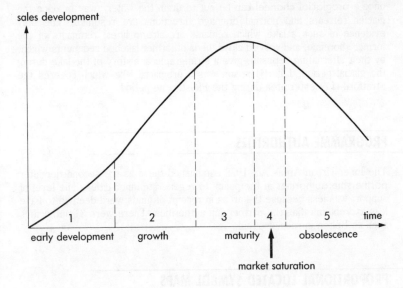

Fig. P.3 The product life-cycle

- the introductory phase of product innovation, e.g. the compact disc;
- the phase of growth when the market for CDs grows as the market for HiFi centres is penetrated;
- the mature phase when sales increase but there is new competition from revamped old products such as digital casettes;
- market saturation (which has yet to come in this example); and
- obsolescence when sales decline and a new product takes over.

Normally new products gain the best image if they are associated with new locations.

PROGLACIAL CHANNEL

◀ Meltwater, Proglacial lake ▶

PROGLACIAL LAKE

A lake formed of glacial **meltwater** which has been trapped between the icefront and other surface features beyond the glacier. The lake disappears once a **proglacial channel** can be cut to drain the water away or once the glacier retreats and normal drainage directions are resumed. The only evidence of such a lake which remains are **strand lines**, remnants of the former shoreline, and **varved clays**, beds of former lakebed sediments which, by their alternating deposits, give a stratigraphical history of the lake during the glacial period. Lake Harrison was a proglacial lake which covered the Stratford–Leicester area during the Pleistocene period.

PROGRAMME AUTHORITIES

The Inner Urban Areas Act 1978 established these as the second tier after **partnership authorities** in the policy to regenerate inner cities. The level of support was less because the areas in receipt of funds were deemed to have lesser problems than the partnership authorities. There were 31 authorities such as Bolton in receipt of aid.

PROPORTIONAL LOCATED SYMBOL MAPS

Maps such as that in Fig. P.4 which use symbols located on the map and are drawn in proportion to the size of the variables or in proportion to the square root or logarithm of the variables.

PROTECTIONISM

A policy adopted by countries and **trade blocs** to give advantage to their own producers of goods and services by excluding the products and services of other countries or trade blocs.

PSAMMOSERE

The sand-dune **ecosystem**, see Fig. P.5. Found in areas with a large tidal range where there is a low gradient beach. There is a **succession** of plant communities on the succession of dunes in which Marram grass becomes the dominant vegetation on the main dunes together with couch grass, bindweed and sea holly. In the slacks a distinct marshland niche develops. Further inland on the stabilised dunes, acid-loving heath plants become common and still further inland, pine trees may seed or be planted. Rabbits have reduced the variety of the species on some dune areas in the past. The ecosystem is particularly vulnerable to human action because dunes are ideal recreational

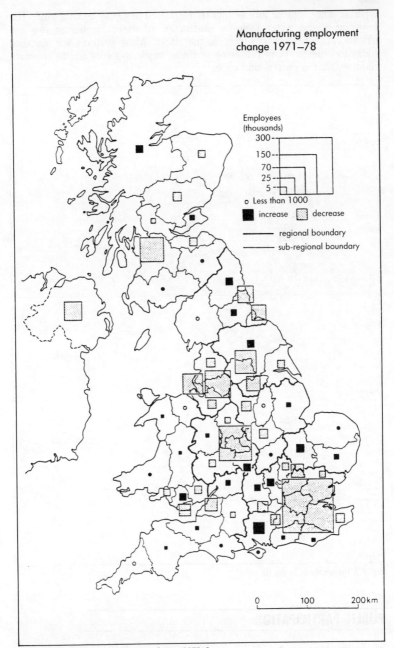

Manufacturing employment change 1971–78

Employees (thousands)
300
150
70
25
5
o Less than 1000
■ increase □ decrease
—— regional boundary
— sub-regional boundary

0   100   200 km

Fig. P.4 Manufacturing employment change 1971–8

environments, may be forested, grazed and used to supply building sand. The lowering of the water table by abstraction of water for human use and irrigation can destroy the niche in the slacks. Most systems are managed today in order to conserve some of these fragile systems and to reinstate them within the coastal landscape.

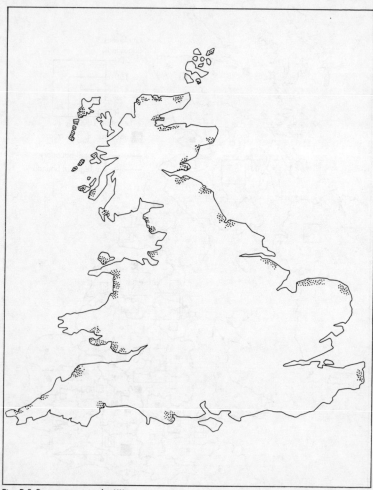

Fig. P.5 Psammoseres in the UK

PUBLIC PARTICIPATION

◀ Town planning ▶

PUMICE

◄ Volcano ►

PUSH MORAINE

◄ Moraine ►

PUSH-PULL MODEL OF MIGRATION

The simplest model of migration which has been merged with the models of **Stouffer** and **Lee** by Carr to produce a tabulation of the factors affecting the decision to migrate.

PYROCLASTIC MATERIAL

◄ Volcano ►

QUADRAT SAMPLING

A method of mapping vegetation patterns by sampling. The vegetation is recorded in each quadrat so that the frequency of plants is known for each sample site which is normally a metre square. The samples may be systematic, i.e. up a slope at regular intervals, or random, i.e. throwing the frame and taking measurements where it lands. The frequency of plant occurence is measured as 'local shoot', the degree of cover, or 'local rooted', the density of the species.

◀ Braun-Blanquet rating system ▶

QUANTITATIVE REVOLUTION

The time in the 1960s when the approach to geographical study changed most radically from a basically regional approach to the introduction of quantitative techniques associated with **spatial analysis** and the **scientific method**. This was the most major shift of **paradigm** to affect geography.

QUATERNARY PERIOD

The latest geological period which comprises the Pleistocene and the Holocene periods. In the early part of the Quaternary the major glacial advances and retreats took place which moulded the landscape of much of northern Europe and indirectly affected the formation of the landscape further afield from the main glacial regions.

QUATERNARY SECTOR

The sector of the economy which relies on high levels of skill and expertise for the transmission and development of ideas. It is based on people, brains and information and comprises research and development (R & D), financial services, education and a range of professional services. Both **producer services** and **consumer services** are part of this and of the **tertiary sector**.

QUOTAS

A specified maximum production for a crop to restrain production. It is a method of controlling output used by the **European Community** as a part of its **Common Agricultural Policy**. Dairy produce has been restricted by quotas.

RAISED BEACH

A former shoreline which has been raised above the present sea level as a result of either **eustatic change** or **isostatic adjustment**. Raised beaches may occur as a succession of levels which gives the geomorphologists clues to the erosional history of an area from both the height of the beaches and the nature of the deposits upon them. Many contain relic features of coastal scenery and may be backed by fossil cliffs which have been degraded by subsequent fluvial erosion.

RANCHING

A form of **sedentary pasturalism** associated with temperate latitudes which relies upon the grazing of large herds of animals for their meat or wool. These **extensive** agricultural practices have intensified output, sometimes as a result of breeding and as a result of improving the grazing and irrigation.

RANKER

Immature soil of limited depth, with a thin A horizon, showing organic matter beginning to mix with unaltered parent material. Lithosol and regosol are other terms used.

RANK–SIZE RULE

This was developed by Zipf in 1949 and is illustrated in Fig. R.1. It states that if the settlements of an area or a country are ranked, then there is a size

relationship to the largest city of the form $Pn = \dfrac{Pi}{n}$, where P = the population of the city ranked nth; Pi is the population of the largest city and n is its rank. When plotted on graph paper the rank–size relationship is a J-shaped curve.

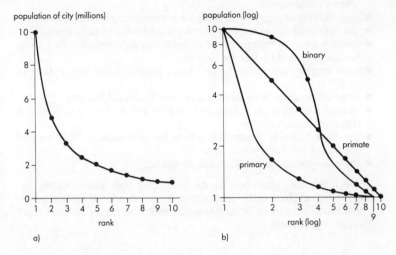

Fig. R.1 Rank-size rule

On Log/Log graph paper a perfect rank–size distribution will form a straight line which is extremely rare. In reality the relationship is primate or binary. Primate relationships are more common in the developing world where the principal city grows out of all proportion to all others in the country. A stepped rank–size pattern is more reminiscent of the pattern found by Christaller.

RATIONALISATION

The reorganisation of manufacturing or service activity to achieve greater efficiency and profit. It may follow a merger, changes in the firm's revenue, or increased competition from overseas. Heavy industry such as iron and steel has rationalised in the face of competition. It can be a stage on the way to deindustrialisation.

RAVENSTEIN'S LAWS OF MIGRATION

The laws were developed on the basis of migration in Great Britain between 1871 and 1881. Ravenstein devised 11 laws:

- the majority of migrants go only a short distance;
- migration proceeds by steps;

- migrants going long distances go by preference to one of the great centres of commerce;
- each current of migration produces a compensating counter-current;
- the natives of towns are less migratory than those of rural areas;
- females are more migratory than males within the kingdom of their birth, but males frequently migrate beyond;
- most migrants are adults; families rarely migrate out of their country of birth;
- large towns grow more by migration than by natural increase;
- migration increases in volume as industries and commerce develop and transport improves;
- the major direction of migration is from the agricultural areas to centres of industry and commerce;
- the major causes of migration are economic.

These laws really apply best to the time when they were written, i.e. during the second phase of the demographic transition in Great Britain. Some, e.g. that towns grow more by migration than natural increase, now only apply to the developing world.

REAFFORESTATION

◀ Forest ▶

RECESSIONAL MORAINE

◀ Moraine ▶

RECESSION LIMB

◀ Hydrograph ▶

RECREATION

Home-based leisure activity which may be active **leisure**, such as sports or passive leisure such as reading. It may also be classified into; formal recreation which is organised, e.g. a visit to a cinema or a day trip in a coach; or informal recreation where the day out is a car trip with no specific destination, or sitting on a beach. Recreation may be based on a resource such as a beach or orientated towards the user such as a theme park.

RECREATION RESOURCE

This is the same as a tourist resource, anything which satisfies demand ranging from a local park or a country park to the beaches of the Mediterranean or the east coast of Malaysia.

RECYCLABLE RESOURCES

◀ Non-renewable resources ▶

RECYCLING

The reuse of non-renewable resources particularly non-ferrous metals so as to conserve the reserves which are rapidly reaching total exhaustion at the final stage of the cycle of resource exploitation. Recycling is also used to conserve energy particularly in those industries where the product uses much energy, e.g. glass. It is of growing importance where there is a need to conserve long term stock or flow resources such as timber for paper pulp. Bottle, paper and aluminium can banks are a prominent feature in West German cities and Munich has over 900 collection points for these recyclable materials.

RED LINING

The term given to a practice employed by mortgage lenders of refusing mortgages to properties in a particular area. The operation of such practices is generally prohibited although it is still possible for the policies adopted by lenders to result in few mortgages being granted in certain areas. Examination of council, bank and building society mortgages have shown different lending patterns.

◀ Green lining ▶

REDUCTION

A form of chemical weathering where the continuous presence of water removes oxygen and converts ferric iron into the more soluble ferrous iron. Its effects can be seen in the blue-grey colouring of gley soils.

REG

A gravel-like desert formed as a result of the wind deflation of the finer particles.

REGION

An area with distinctive human and physical unity clearly distinguishable from other surrounding areas. Regionality may be based on a range of criteria although regions based on a multiplicity of criteria are less easily distinguished. This concept which was at the heart of geographical studies, has received recent re-evaluation particularly in the light of the National Curriculum which expects some form of area study. The most commonly accepted division of regions is into **formal** and **functional**.
◀ Friedmann's centre-periphery model ▶

REGIONAL GEOGRAPHY

The study of areas on the earth's surface with the aim of examining and explaining the inter-relationships between the various aspects of the physical and human environments of the area. This form of study was the *raison d'etre* of geography until the 1950s although it declined in popularity during the **quantitative revolution**. More recently it has seen a resurgence because many of the aspects of modern geographical study especially within the fields of **welfare geography** and **humanistic geography** are based on the inequalities that can be observed between areas. In 1990 the National Curriculum to be studied by all 5–16 year olds in Great Britain reasserted the view that all should receive some training in regional geography, especially that which enables people to know where places are. Therefore there is a return to some of the more old-fashioned aspects of regional geography nick-named 'capes and bays'. At 'A' level regions continue to be used as a framework for the analysis of a particular process or an issue or problem.

REGIONAL PLANNING

This was developed in Britain in the 1930s as a response to regions suffering economic problems and, by the 1940s, had also acquired a concern for the urban environment. It concentrated almost exclusively on industrial policy from the 1940s until the 1970s when office location was belatedly added to the brief. Since 1979 regional planning has been almost totally dismantled except for the purpose of obtaining grant aid from the **European Community**. Regional planning is still a strong element in French national plans. In Italy the

Mezzogiorno scheme has been a major economic planning scheme for two decades. Many question whether regional planning does anything other than formalise the processes which are already operating in regions. It is often seen as a further bureaucratic tier in government which is unneccessary and expensive.

REGOLITH

The zone between the rocks of the earth's crust and the surface of the soil. It is therefore soil and partially disintegrated rock together with other deposits such as windblown sands. It is the zone in which most **weathering** takes place and where all the **pedogenic** processes take place.

REGRESSIVE POPULATION PYRAMID

◀ Age-sex pyramid ▶

REILLY'S BREAK POINT THEORY

◀ Sphere of influence ▶

REMOTE SENSING

The interpretation of phenomena on the earth's surface from the evidence provided by photographs or **images**, taken by satellites orbiting the earth. The images are used to map phenomena such as vegetation, land use, weather patterns, heat and pollution. The images are false colours which, when interpreted through the analysis of the 'pixel' colours, provide very detailed information on the changes in, e.g. landuse through the seasons.

RENDZINA

A type of intra zonal soil asociated with highly calcareous materials. It has an AC profile where the A is characterised by a brown or black friable surface layer underlain by a grey to pale yellow calcareous material.

RENEWABLE RESOURCES

Resources which are continually available for use. They may be used without any fear of their exhaustion because they are replenished sufficiently rapidly for their use to have little effect on the stock. Renewable resources are subdivided into **stock** or **flow resources** which can be depleted by human use as well as be increased by human activity. Stock resources can be short term like a crop of potatoes, medium term such as water or long term such as a forest. The greatest threats from human activity obviously are to the long term stock resources such as the Amazonian forests. Non-stockable or **continuous resources** are always available, e.g. wind power.

REPATRIATION

The forced **return migration** of people to their country of origin. It occurs as a result of;

- declining needs for immigrant labour which cannot afford to go back home,
- the return of peoples who are not deemed to have fulfilled the requirements to be either political refugees or asylum seekers, e.g. the Vietnamese boat people in Hong Kong in 1989–90.

RESEARCH AND DEVELOPMENT (R & D)

A **quaternary activity** involved in the innovation of new economic activity which has grown strongly in the **developed world** in recent years. It has its own locational requirements which are related more to a pleasant working and living environment rather than any other locational requirements. Antipolis near Nice on the French Riveria is one such environment.

RESERVES

The amounts of a resource available under the current technological and economic conditions. They change as technology changes and the demands of the economy and society change. Oil reserves have risen as more sophisticated exploration techniques discovered more fields in formerly inexpoloitable locations. However, oil reserves fall once society makes more use of oil and its by-products.

RESIDENTIARY ACTIVITIES

◄ Non-basic activities ►

RESOURCES

◄ Capital resource, Human resource, Natural resource, Non-renewable resource, Renewable resource ►

RETIREMENT MIGRATION

The movement of people before, at, or after retirement. It has generally been towards coastal and rural areas in the United Kingdom. Some lack the means to move and so they, retire *in situ*. Retirement migration results in concentrations of the elderly which can put pressure on the local social services in the future once the healthy old become more infirm.

RETURN MIGRATION

The movement back to the area of origin by labour migrants who either return voluntarily having satisfied their needs, or as result of repatriation schemes which assist their return.

RE–URBANISATION

The movement of people and jobs back to the city and especially to those areas which were abandoned during the processes of suburbanisation and counter–urbanisation. It involves the recolonisation of areas often left derelict by deindustrialisation. The new uses normally contrast markedly with the former uses so that middle class homes replace those of the workers (e.g. Wapping) or are built on former industrial land (e.g. Beckton gas works, Salford Quays). The new activities are focused upon office work in high technology firms, e.g. NEC at Salford Quays or extensions of city centre tertiary activities, e.g. Canary Wharf. The process is only partly synonymous with gentrification.

RHOURD

A star-shaped sand mound up to 1km across and 150m high formed by winds from different directions in a desert.

RIA

A coastal valley which has been submerged as a result of rises in sea level. The valley was formed during periods of lower sea level, i.e. the Pleistocene and eustatic change has brought about the flooding of the valley.

RIBBON LAKE

A lake which has been formed in a glaciated valley often by deposits acting as dam to the overdeepened part which becomes the lake basin, e.g. some of the lakes in the Lake District. Glacial lakes have a tourist potential especially in upland areas.

RICHTER SCALE

◀ Earthquake ▶

RIFFLE

A depositional feature on the floor of a river channel where the velocity in the meander is less. They tend to move towards the inner side of the meander where they join up with the point bars. It is thought that the occurrence of riffles and pools is linked to the formation of meanders.

RILL

A small temporary channel down which water flows mainly during periods of intense precipitation when the infiltration rate is exceeded. They develop especially in unvegetated areas where sheetwash and sheetflow are also common and result in erosion by the many small channels which carry the water. Rills will coalesce in time to form gullies. Rills can be seen on ploughed fields especially while they are being prepared for the sowing of winter wheat when the surface has been compacted by machinery and by harrowing.

RILL EROSION

An erosion process associated with confined flow in rills.

RISING LIMB

◀ Hydrograph ▶

RIVER POLLUTION

This is a people-made **environmental hazard** produced from a variety of sources. Effluent from sewage works and raw sewage from sewage outfalls may cause river pollution as many industrial spillages such as the Sandoz disaster on the Rhine at Basel in 1986 which sent toxic chemicals into the river destroying wildlife down its length. Nitrate fertilisers drain via the **groundwater** and flow directly as **surface flow** into streams causing problems for people who may drink that water once it has been taken into the water supply. In 1988 aluminium was found in the drinking water in Camborne and this, the result of an accident to a river water supply, caused health problems among the population.

ROCHE MOUTONÉE

An assymetrical rock hill in a glaciated landscape which has been shaped by the passage of a glacier. The upstream side is rounded with striations, scratch marks made by **abrasion**, while the downstream side is steeper due to **plucking**.

ROCK STEP

An irregularity in the long profile of a glacial valley caused by an outcrop of a more resistant rock (riegel) or by the extra erosive power of two glaciers below a confluence.

RORO

The acronym for roll-on roll-off vessels used to transport cars and lorries across major sea channels such as the English Channel. It is a modern form of **piggy back system**.

ROSTOW'S STAGES OF ECONOMIC GROWTH MODEL

A country passes through five stages of growth on its way to economic maturity, see Fig. R.2:

- traditional society which is dominated by subsistence agriculture;
- pre-conditions for take-off where trade expands and new inventions and technologies are introduced;

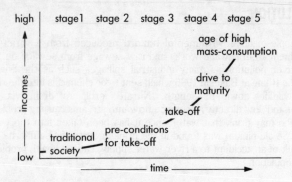

Fig. R.2 The stages of economic growth model (after Rostow)

- take-off, the key stage when with increased investment modern industrial methods of production replace traditional ones and growth is self-sustaining;
- drive to maturity during which a more complex economic system evolves in an urban society; and
- the age of **mass production and consumption** in which there is an increasing shift towards consumer goods and the welfare state.

The model was based on the evolution of the European and North American economies over the past 200 years and has been regarded as time specific. Its applicability to the present day has been questioned because it is **eurocentric**. It assumes that all states pass through the sequence. It overemphasises capital's role in development when some countries such as Tanzania are trying a different, socialist form of development. There is also scant attention paid to non-economic aspects of development. It does not take into account the movement of aid which might be dispersed more in line with political dogma than the need to help countries develop.

ROTATION

The growing of different crops each year on a field to assist in the natural return of nutrients to the soil and to help prevent **soil erosion**. The **three field system** was an early, inefficient form of rotation. Fertilisers removed some of the need to have rotation although many see it today as a more diligent and environmentally friendly way of conserving the soil resource.

ROTATIONAL SLIP

A major process of **mass movement** clasically associated with permeable rocks overlying impermeable rocks. The angle of dip of the strata and the presence of water which acts as a lubricant aid movement along a **slide plane**. Some of the best examples of slips occur on marine cliffs such as those at Barton-on-Sea, Hants and Ventnor where the whole undercliff area has been formed from some large landslips. In these cases the faster removal of debris at the foot of the slip has the effect of accelerating the process.

◄ Slumping ►

RUNNEL

A small depression parallel to the shoreline lying between sand ridges on a **beach**. It is also called a swale.

RUNNING MEAN

Sometimes called the 'moving average', it is calculated by taking the **arithmetic mean** of overlapping data sets over a period of time in order to establish trends in the values. Data could be collected for successive, overlapping 5 year periods, e.g. 1980–5, 1981–6, 1982–7 etc.

RURAL DEPOPULATION

The gradual decline of population in a rural area occasioned by outmigration. It can also be caused by natural decrease which is not balanced by **immigration**. Job loss and mechanisation in agriculture, the perceived benefits of urban life, poor rural housing and lack of facilities have all encouraged depopulation. Many also leave to get married. In the **developing world** many migrate to find work – **labour migration**. Depopulation has generally been counterbalanced in all but the remotest areas by repopulation caused by **counter–urbanisation**. Where it does occur, property is abandoned possibly to be used as **second homes**, services are lost and people have to travel further to obtain services, the population ages and, in the developing world, labour is lost to the urban areas.

RURAL DEVELOPMENT PROGRAMMES

Used by developing countries to assist in economic development. The aim is to increase food production and so save on the costs of importing food; the **green revolution** has been the best known strategy for improving food output. Other strategies have involved the increase in export crops such as oil palm and pineapples in Malaysia, the reformation of the land holdings to provide larger holdings capable of being worked by mechanical equipment and infrastructural developments such as the building of roads, laying of water pipes and waste disposal systems. Land reform programmes to remove absentee landlords of vast estates have been attempted in, e.g. Peru. In Cuba and Tanzania land reform has been more radical and has involved **state farms** and **cooperatives** respectively.

RURAL GEOGRAPHY

The study of the geographical aspects of human activities in rural (non-urban) areas. It normally refers to the study of agriculture, forestry, population and settlement and associated aspects of **physical geography**, especially **soils**. Because it is difficult to separate urban population and settlement from rural, the term is not used much except for the study of the economic aspects of the countryside.

RURAL–URBAN MIGRATION

See **rural depopulation,** although not all of this migration necessarily represents depopulation. On the whole it is more likely to be a movement of the younger elements of the population towards the perceived opportunities of urban life.

RUSTBELT

The North American term given to the areas of declining industry and **deindustrialisation**. It is synonymous with the concept of **sunset industrial** areas.

SALINATION

The process by which salts accumulate with the soil matrix.

SALTATION

A method by which sediment is transported by wind and by streams. The particles bounce or hop along because they are too heavy to remain in suspension. The size of particles which are moved depends on the velocity or speed of the current of water or air.

SAND RIPPLES

Small undulations in the sediments produced by the waves on gently shelving beaches.

SAND SEA

An area where sand in desert regions accumulates in the form of large expanses of dunes such as are found in parts of the Sahara. It is also known as an erg.

SATISFICERS

A general term which refers to the behaviour of decision makers when faced with a choice and their knowledge on which to base their decision is imperfect. Therefore they make the choice which best suits them. Thus a farmer/landowner may choose to keep a horse for family recreation rather than grow a crop which would maximise the return on the land. Similarly a shopper will choose to journey further because the shops in the smaller market town are perceived to be better quality. In both cases the psychic value outweighs the rational maximising behaviour.

SCATTER GRAPHS

A useful method of comparing data to see if there is any association or correlation between the two variables prior to statistical testing. The axes are

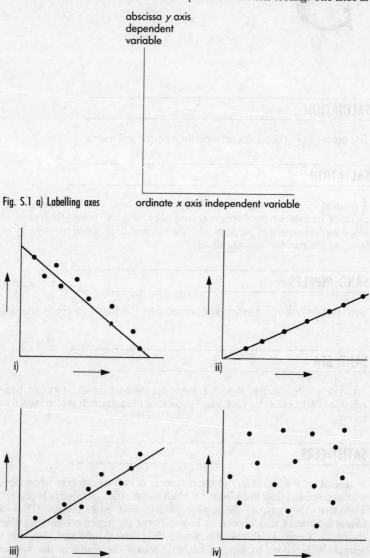

abscissa *y* axis
dependent
variable

Fig. S.1 a) Labelling axes   ordinate *x* axis independent variable

i)   ii)

iii)   iv)

Fig. S.1 b) Scattergraph patterns

labelled as in Fig. S.1a; Fig. S.1b shows the following patterns; i) is an example of a scatter graph which shows a negative correlation; ii) shows a perfect correlation; iii) a positive correlation and iv) no correlation.

SCIENTIFIC METHOD

An approach to learning which progresses through the definition of an **hypothesis**, the collection of relevant data and the processing of that data to test the hypothesis. The interpretation of the results then enables the researcher to confirm or deny the original hypothesis and, if necessary, define a new hypothesis. If the result can be generalised across a wide variety of cases, it may be formulated into a law. It has become the normal method of research organisation especially within **physical geography** and is a good basis for developing a field project at A-level.

SCORIA

◄ Volcano ►

SEA LEVEL CHANGES

◄ Eustatic change, Isostatic adjustment ►

SEA WALL

A construction running parallel to the shoreline which is designed to prevent the erosion of the shoreline. Some sea walls may run offshore in order to create an artificial harbour, e.g. Brighton Marina. They may be constructed to prevent catastrophic flooding of low lying coastal areas such as the Essex coast; in the Netherlands they are called sea **dykes**.

SECONDARY DATA

Data obtained from published sources. The population census and meteorological records are examples of secondary data used by geographers.

SECONDARY ENERGY

Energy derived from products which have been converted to enable them to be used. Electricity derived from thermal, oil, gas and nuclear power stations is the major source of secondary energy.

SECONDARY SECTOR

The sector of the economy that processes **primary products** and other products of the secondary sector into other manufactured goods. It may be called the industrial sector.

SECONDARY TOURIST RESOURCES

The facilities such as cafés, bars and fun-fairs which help the visitor to enjoy the visit to a primary tourist resource.

SECOND HOMES

Dwellings normally, but not exclusively, in the countryside whose residents live elsewhere for the greater part of the year. These can be holiday retreats, purpose built settlements such as La Grand Motte in Languedoc or in villages which were abandoned by **rural depopulation** only to be recolonised by second home owners. Time share apartments are another form of second home where the resident only owns the right to ocupy the dwelling for a set period of time each year. Second home ownership might deprive many of the local population of the opportunity to own a home. In some areas, e.g. Wales, this has led to violent protest.

SECOND WORLD

The countries of the socialist world in the **developed world** where the state ownership of the means of production and distribution is the dominant economic force. Developments in East Europe in 1989 will necessitate a rethink of this definition. Not all socialist states are in the second world; many such as Vietnam, Tanzania and China are part of the **developing world**.

SECTOR MODEL

◄ Hoyt ►

SEDENTARY PASTURALISM

Animal husbandry on **extensive** farms which takes the form of the 'hamburger farms' in Rondonia, Brazil. **Ranching** in the temperate areas is also a form of sedentary pasturalism.

◄ Intensive subsistence, Sedentary agriculture ►

SEGREGATION

The separation of groups of people from one another on the basis of their class, socio-economic group, religion or race. Normally geographers study residential segregation based on the way in which income, social status and stage in the life cycle interact to influence our choice and location of home. At its most extreme, segregation becomes **apartheid** and is founded in law. Other cultural differences such as language (Flemings and Walloons in Belgium) and religion (Roman Catholics and Protestants in Northern Ireland) also lead to segregation.

◀ Apartheid ▶

SEIFS

Longitudinal dunes found in the **sand sea** or **erg** deserts. The dune can be several kilometres in length and their origin is uncertain. Some think that they form in the lee of an obstacle and others think that they are the horns of a **barchan** which have been split by a blow-out of the centre of the crescent dune. These dunes are parallel to the wind direction especially in the Australian desert where they extend for up to 100km.

SEISMIC ENERGY

◀ Earthquake ▶

SERVICE INDUSTRY

A general term used to cover the whole range of **tertiary** and **quaternary** economic activities.

SET-ASIDE POLICY

A policy introduced in the **European Community** to combat the overproduction of agricultural products. Farmers are paid not to grow crops and to leave land fallow. It is essential to provide the farmer with some income for the land for social and political reasons.

SETTLEMENT FORM OR SHAPE

The outline and internal shape of rural settlements such as *linear, nucleated* and *green*. 'Form' is also the term used to described the morphology of an urban area.

SETTLEMENT GEOGRAPHY

The study of human settlements in both rural and urban areas, their growth, functions and patterns. **Urban geography**, a branch of settlement geography, is one of the largest fields of geographical enquiry today and itself contains many fields of research and study.

SETTLEMENT PATTERN

The arrangement of settlements on the Earth's surface. The term is normally associated with rural settlements and describes the degree of nucleation or **dispersion** and the actual shape or **settlement form** of the nucleations.

SHEAR STRENGTH

The resistance of a material to forces applied to it. It often refers to the ability of a soil to resist the forces of **gravity** and not move downslope under the forces of **solifluction** and **creep**. The presence of water reduces the strength of the soil to resist the forces and so at a certain point shear strength is exceeded and material slowly slides downslope.

SHEET FLOOD

The method by which large amounts of soil and debris are removed from the surface by **overland flow**. Sheet erosion which is the same as sheet flood is the combined product of rainsplash erosion and the flood which removes the soil particles. It is very common on areas of exposed soil after ploughing and will result in the accumulation of silt at the foot of a slope against a hedge or wall.

SHEETFLOW

A process of **mass movement** which is found on unvegetated slopes. It is the movement downslope of surface material by **overland flow** involving the shifting of the material by raindrop impact and the carrying of the particles by the water flowing down the slope or **sheetwash**.

SHEETING

A form of **mechanical weathering** normally associated with igneous rocks where the upper layers of the rock split off.

SHEETWASH

A thin film of overland flow water over all parts of the slope. The removal of material by this flow is called sheet erosion.

SHIFTING CULTIVATION

Cultivation practised by a semi-nomadic group of people who cultivate a small area of land until it is exhausted. Very often the nutrient levels are raised by burning the vegetation so that the nutrients released from the ashes can be used by the crops. Once the land has been exhausted the group moves on to a new area. It is regarded as **extensive agriculture** because a large area of land is needed to support a group of shifting cultivators over a period of years.

SHIFTING SAND

Sand movement which may bury agricultural land or destroy an **oasis**. This **environmental hazard** has threatened fertile areas in the Netherlands which border the sandy heathlands of the Veluwe. Here the dunes have been blown onto the fertile land so reducing the field size and preserving fossil soils beneath the sand.

SHIMBEL INDEX

A measure of accessibility used in **network theory**. It measures the total number of edges needed to connect any vertex to all others in the network by the shortest path. The lower the value the greater the accessibility and centrality of the place.

SILL

1 A horizontal **volcanic intrusion** which has been forced between bedding planes. Where sills come to the surface and form a resistant capping they can provide very pronounced steep slopes which dominate a landscape, e.g. Salisbury Crags in Edinburgh.
2 The ridge across the mouth of a **fjord**.

SIMPLIFICATION

The reduction of plant communities in a habitat so that people may produce more of one particular plant from the community. Agriculture is a form of ecosystem simplification.

SINCLAIR'S MODEL

The value of the land for agriculture close to the city margin is lower than in areas more distant from the city. The evidence is that low intensity uses such as the keeping of large pets creeps into the fringe land. Low value is also caused by damage to crops by vandalism and the speculative holding of land in anticipation of development. Efficient transport enables intensive uses to locate further away from the market.

SINK HOLE

The widened area where two **grykes** meet, and where water can penetrate the **underground cave** system in an area of **carboniferous limestone**. It should not be confused with a **swallow hole**.

SITES OF SPECIAL SCIENTIFIC INTEREST (SSSIs)

Areas designated as a result of the provisions of the National Parks and Access to the Countryside Act 1949 and the Wildlife and Countryside Act 1981. There are 4000 SSSIs where flora and fauna are conserved by agreement between land owners, occupiers and the local council. The Nature Conservancy Council establishes the sites.

SLIDE

A process of **mass movement** in which the soil, rock and debris move downslope along a bedding plane or an erosional surface lubricated by water.

The 1963 Vaiont Dam tragedy was caused by a slide entering a reservoir. Water is the main lubricant and most slides occur after periods of heavy precipitation.

SLIDE PLANE

◀ Rotational slip ▶

SLOPE

The fundamental unit of all landforms whose form is the product of the processes operating on it. It is measured in terms of its angle, profile and the depth of the **regolith** at points down the slope. The form of the slope is the product of past processes and this in its turn influences the present processes which operate on it. Material moves down a slope under the influence of **gravity** although water is essential for most of the processes of **mass movement** to operate. Slopes are altered by rainsplash, **sheetwash** and movement in **solution**. There are **models** of slope retreat.

SLOPE FORM

There are four basic units to a slope as shown in Fig. S.2: the **waxing** or

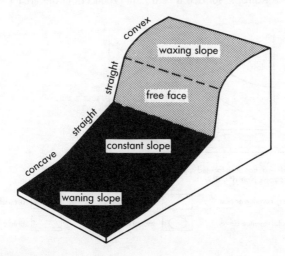

Fig. S.2 The slope profile

convex slope; the free-face; the constant slope and the waning or concave slope. The talusfoot is a fifth unit which some geomorphologists place between the constant and waning slopes.

SLOPE SYSTEM

A sub-system of the catchment basin system in which the processes of slope decline take place.

SLUMPING

A form of mass movement on a slope where the coherence of the surface layers is lost due to saturation of the regolith. The material then moves along a slide plane which might be lithological and often takes the form of a rotational slip. It is a more general term and encompasses slips, flows and other gravitational earthflows.

SMITH'S SPATIAL MARGINS MODEL 1966

This model of industrial location, shown in Fig. S.3, plots both revenue and costs against distance to show that profits can be obtained over a broad area Y – Y^1 on the isotropic surface $d^1 – d^2$. All producers in the area Y – Y^1 will

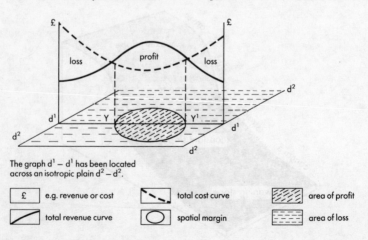

The graph $d^1 – d^1$ has been located across an isotropic plain $d^2 – d^2$.

| £ | e.g. revenue or cost | total cost curve | area of profit |
| total revenue curve | | spatial margin | area of loss |

Fig. S.3 Smith's spatial margins model

obtain a profit. This model is an improvement on earlier theories in that it pays much more recognition to the concept of the **satisficer** in industrial location and acknowledges the imperfections of locational knowledge among decision makers.

SOCIALIST FARMING SYSTEMS

Farming organisations such as **collective farms, state farms** and some forms of **co-operative farming** which have been established by the socialist states as a part of their organisation of the means of production. Normally output is planned and, as such, does represent a form of state control of the system. Such planning has not been successful mainly because of unrealistic expectations which took no account of climatic variability or of the lack of incentives in the organisation of agriculture.

SOCIO-ECONOMIC GROUP

A classification based mainly on the occupation of the head of household which was developed for use in the UK census and other official surveys. There are 17 groupings and all jobs are placed into one grouping. It is a more accurate definition than the rather vague class definitions used by geographers: wherever possible you should use socio-economic group data.

SOCIO-POLITICAL ENVIRONMENT

The factors such as government legislation or the lack of it, taxation, regional policy, urban policy and subsidies which may have an impact on **locational decision making**.

SOIL

The unconsolidated mineral material on the immediate surface of the earth that has been subjected to the influence of certain environmental factors. These include climate, parent material, topography, time and micro/macro faunal activity. A soil is produced which differs chemically and physically from the parent material. It now acts as a natural medium for plant growth.

SOIL ACIDITY

This measures the activity of the hydrogen ion in the liquid phase of the soil system. It is measured and expressed as a **pH** value.

SOIL COLOUR

Described using the Munsell colour system. The colour of the soils is compared with a set of standard colours. In this way three simple variables of colour can be obtained: hue, value and chroma. For example, 10YR 6/4 has hue = 10YR, value = 6 and chroma = 4.

SOIL EROSION

The detachment and movement of soil particles by agents of erosion; water flow, raindrops, wind, etc. Many types of soil erosion therefore exist; gully erosion, rill erosion, rainsplash erosion, sheet erosion, etc.

SOIL HORIZON

A layer of soil or soil material approximately parallel to the land surface and recognizably different from the layers above or below in terms of its physical, chemical or biological properties. The horizons are different as a result of either additions, subtractions or more intense alterations than those adjacent layers. The horizons are normally designated by a letter code and additional subscripts may be used to define specific (often called diagnostic) horizons within them.

- O horizons: Organic horizons, sub-divided depending on the intensity of decomposition.
- A horizons: Horizons showing the intimate mixing of both the organic and inorganic components of the soil. Ap = ploughed A horizon.
- B horizons: Mineral layers characterised by additions or alterations.
  B_s Horizon enriched by Fe and Al. Sesquioxide rich. *Spodic* horizon.
  B_t Horizon enriched with clay-sized material, seen as skins or coats around ped surfaces. *Acrillic* horizon.
  (B) Horizon of alteration where all evidene of rock structure has been destroyed. *Cambic* horizon.
  B_g Horizon of alteration characterised by Fe in its reduced form resulting from water-logging. Gleyic properties.
- E horizons: Mineral layers of depletion.
  E_a Depleted of sesquioxides. *Albic* horizon.
  E_b Depleted in clay-sized material.

SOIL PIT

A pit dug in the field for the investigation and description of soil profiles.

SOIL STRUCTURE

The combination or arrangement of soil particles into larger secondary particles, units, **peds** or aggregates. These are the natural units that the soil breaks into and are characterised and classified according to their shape, size and stability.

SOIL TEXTURE

The relative proportions of individual inorganic particles of various sizes. These various proportions define the soil texture or textural class, for example, loamy sand.

SOIL WATER

Soil water is held in the soil as a result of a number of forces: these result from the presence of the soil solid phase, gravitational field, dissolved salts and external gas pressure. Each one has an effect on the total force, which is called **total potential**.

SOLAR POWER

◀ Energy resources ▶

SOLIFLUCTION

A process of mass wasting commonly associated with **periglacial** areas. It is the downslope movement of saturated soil caused by the melting of surface areas in summer causing the soil to become saturated and easily influenced by gravitational forces. The soil loses its **shear strength** due to the presence of the water and the material moves down slope. It can result in the formation of solifluction sheets and solifluction lobes. Vegetation will slow the process down.

SOLONISATION

Process leading to a solonetz soil, i.e. a soil dominated by sodium carbonate. It is highly alkaline, with some leaching of soluble salts, and has a grey or brown B horizon sometimes enriched with clay which contrasts with the black surface crust that forms at the surface during dry conditions.

SOLUTION

A method by which material is transported by streams. More importantly it is the process by which a material is transformed from a solid into a liquid state by combining with a solvent chemical. It is a process of **chemical weathering** which dissolves rocks such as **carboniferous limestone** by slightly acidic rainwater.

SOLUTION LOAD

That part of a stream's **load** carried in solution.

SORTING

A process occurring during transport of eroded sediments by which particles are graded and deposited by size. Within a delta deposits which are heaviest are deposited first and the finer sediments are carried further out to sea. It is very evident in **periglacial** conditions such as **patterned ground** and on scree slopes.

SPATIAL ANALYSIS

The approach to the study of geography which resulted from the **quantitative revolution**. Studies emphasise the interdependence of phenomena and the regularities in the patterns and processes of the phenomena and depend on the quantitative analysis of the phenomena in order to achieve rigour. As a result spatial analysis has been criticised because it contains no recognition of **behavioural** phenomena.

SPATIAL DIVISION OF LABOUR

The regional division of dominant types of employment in a country which results from processes of locational change and economic development. It normally results from the concentration of control in the dominant, core region. In the UK the core region is London and the South East whereas routine production is carried out in the North and West where there is very little employment in the control function. As a result high level skills and managerial and financial employment concentrated in one part of the country and industrial work of a more routine nature in the other areas. A further development of this concept is the **international division of labour**.

SPEARMAN'S RANK CORRELATION COEFFICIENT

A **non-parametric** test which tests the relationship between two sets of ranked, **ordinal data**. The formula is as follows:

$R_s = 1 - \dfrac{6\Sigma d^2}{n^3 - n}$ where d = equals the difference between each pair of ranks

n = the number of pairs of ranks.

If the result is $+1$ then there is total agreement between the two data sets, whereas if the result is -1 the rise in one set of data results in the fall in the other.

SPECIAL AREAS

◀ Development areas ▶

SPECIAL AREAS ACT 1934

The original legislative base for regional planning in Britain which granted assistance to the North-East, South Wales, West Cumberland and Clydeside/North Lanark.

SPECIAL DEVELOPMENT AREAS

These were established in 1964 to provide additional support to areas which had, or were likely to have persistently high unemployment. They received the highest levels of aid for new economic activities. These areas were abolished in 1984.

◀ Development area, Intermediate areas ▶

SPHERE OF INFLUENCE

The area over which an urban centre delivers its services. It is predicted by using **Reilly's break point theory**.

SPILLS

A general term used to describe the accidental loss of oil and/or chemicals from tankers, pipelines or industrial plants either onto land and into the system or into the aquatic system. Oil destroys the natural insulation and bouyancy in birds and mammals, kills shellfish and so disrupts the foodchain causing a loss of livelihood for coastal communities. Other spills can ruin the tourist potential of a resort albeit temporarily, kill life in rivers and threaten water supplies.

SPIT

A long narrowing deposit of sand and shingle which extends either out to sea or into an estuary from a coastline. It is produced by **longshore drift** where the coastline suddenly changes direction in an area of small tidal range. They build up as a series of lateral ridges out into the water which have a recurved or hooked end. There are several examples of spits along the Hampshire coast between Hurst Castle and Langstone Harbour at the entrance to all the harbours and estuaries. A cuspate foreland is formed when two spits meet as at Dungeness. A spit which links an island to the mainland is called a tombolo. Others block an estuary to form a *bar* which then encloses a freshwater lagoon. Along the Baltic coast of the German Democratic Republic the spits or *nehrung* enclose coastal lagoons called *haffs*. Saltmarshes are often found in the lee of spits and develop a very distinctive ecosystem. Spits are favoured recreational areas on the coast and their overuse threatens their continued existence both as landforms and ecosystems.

SPONTANEOUS SETTLEMENT

Also known by terms such as shanty town, squatter settlement, favela, barriadas and bustee. They are rudimentary areas of housing without sanitation, water supply or any infrastructure. At worst they are a drainage pipe or a blanket in a gutter and at best 'slums of hope', i.e. basic housing of an increasingly permanent nature. The alternatives are camps, **new towns** or site and service schemes. The latter provide a foundation and sanitation but leave the dwellers to provide the rest as and when they can afford it. These settlements are normally associated with urbanisation in the **developing world**, but they are found in the **developed world**, e.g. in Paris in the 1950s and the 'cardboard city' which has grown up around the southern end of Waterloo Bridge in London, housing the young homeless and the down and outs.

SPREAD EFFECTS

◄ Growth pole theory ►

STABILITY

Air density depends on temperature with cold air being denser than warm air. If rising air remains colder than the air above it the atmosphere is described as being stable, since the rising air will be denser and hence buoyancy will tend to force it back down. Dry air cools at the DALR as it rises, so an atmosphere with an environmental lapse rate (ELR) greater than the DALR is said to be absolutely unstable, because any rising air parcel would be surrounded by air at a lower temperature and buoyancy would tend to accelerate the air parcel upwards. If the ELR is equal to the DALR the atmosphere is said to be neutral as any air parcel rising up through the atmosphere would be surrounded by air at the same temperature. However, such an atmosphere is conditionally unstable for if the rising air cooled to its dew point temperature the cooling on any further ascent would be at the slower SALR. The air would then be warmer than its surrounding and would accelerate upwards. Finally if the ELR is less than the SALR the air is said to be absolutely stable as any rising air whether saturated or unsaturated will always be colder and hence denser than its surrounding.

Air can, however, be forced to rise even if the atmosphere is stable such as when air is forced to rise over mountains or when warmer stable tropical air is forced upwards along a warm front.

STACK

A single column of rock standing in the sea after marine erosion has removed the surrounding rock. Marine erosion has exploited lithological weaknesses by differential erosion and often the stack is the remnant of an arch. The Needles on the Isle of Wight are a series of stacks. See Fig. A.7.

STADIAL MORAINE

◀ Moraine ▶

STATE FARMS

A socialist agricultural system which is distinguished by its organisation and not its output which will be the same as any other farm in similar environmental conditions. The farm is a state-run business and all the workers are employees of the state. The profits go to the state and not to the workers. It is found in E. Europe, the USSR and China although the events of 1989–90 might see the abolition of this symbol of Stalinist thinking. State farms also exist in Cuba and Tanzania where the output and the organisation have been modified to suit the particular conditions of these two countries.

STEMFLOW

◄ Throughfall ►

STEPPE GRASSLAND

◄ Temperate grassland biome ►

STOCK RESOURCES

◄ Renewable resources ►

STOCKS

The total components both discovered and undiscovered which go to make up the earth. Some of these are unusable because they are inaccessible, e.g. iron in the earth's core.

STONE CIRCLES

◄ Patterned ground ►

STONE NETS

◄ Patterned ground ►

STONE PAVEMENT

Small stones which remain after the removal of finer grained material by the wind in a desert environment. It is often known as a **desert pavement**.

STONE POLYGONS

◀ Patterned ground ▶

STOPOVER

The term used by airlines for passengers who interrupt a journey to stay at an intermediate point. Singapore is a major stopover point on the routes from Europe to Australia and has developed a tourist industry on this basis.

STORM BEACH

The accumulation of coarser sediments which are found at the top of a beach which have been deposited there during storm conditions when the waves are strong enough to carry the shingle well above the normal levels.

STOUFFER'S INTERVENING OPPORTUNITIES

This theory relates the number of migrants directly to the number of opportunities and inversely to the number of **intervening obstacles**. The opportunities are housing, jobs and environmental factors.

STRAND LINE

Former shore lines which indicated sea levels which were higher in the past especially in areas affected by **isostatic** readjustment after the removal of the ice sheets.

◀ Proglacial lake ▶

STRATIFICATION

1 The arrangement of the vegetation in layers which is measured as a part of vegetation mapping.
2 The arrangement of the soil in layers.

STREAM FLOOD

◀ Flash flood ▶

STREAM ORDER

A method of classifying streams within a drainage system so that the initial streams within a basin are first order and where two of these meet then becomes a second order and when this meets another second order stream it then becomes a third order stream. If a first order stream joins a third order stream there is no increase in the order number at that confluence. The basin can be analysed and compared with other basins in terms of the number of streams per order and the average length of each stream order.

STRUCTURALISM

An approach to the study of human geography which seeks explanations of the patterns which are observed in the broad social structures that underpin society. Geographers have tended to show how the distribution of many facilities in urban areas may only be explained by reference to the nature of our capitalist society with its profit motive. In contrast the socialist structure of East European society is the process which best explains the patterns of economy and society in, e.g. Poland. **Marxist geography** adopts a structuralist approach.

STRUCTURE OF INDUSTRY APPROACH

An approach to the study of industrial location which assumes that firms are not single product enterprises, which was the assumption of **Weber, Losch, Hoover, Hotelling** and others. It recognises the fact that they are multi-product enterprises receiving raw materials from both **primary producers** and other manufacturing firms and selling to many markets. Companies also have shareholders and share-holding workers rather than a single owner. Firms have more than one plant with a headquarters, main plant and branch plants all of which are subject to different locational forces.

STRUCTURE PLAN

◄ Town planning ►

STUBBLE BURNING

A practice of burning the straw and stubble after the grain harvest which has been increasingly practised in Great Britain. Burning disposes of unwanted straw more rapidly than ploughing in because the straw is slow to decompose. The practice may lead to soil erosion after burning and is now regarded as a form of atmospheric **pollution**. As a result the practice will be discontinued in the 1990s. Alternative uses are being found for the straw, e.g. as fuel for special heating furnaces which are being installed in large country residences.

SUB-CLIMAX

A plant community that has been stopped in its development towards a climax by physical factors. In contrast to the **subsere** the sub-climax is regarded as a permanent halt in the plant community's evolution.

SUBDUCTION ZONE

The area along the edge of a continental plate where the oceanic plate is being destroyed as it is pushed beneath the lighter continental plate at the **destructive margin**. The zone is marked by an **oceanic trench** and extends

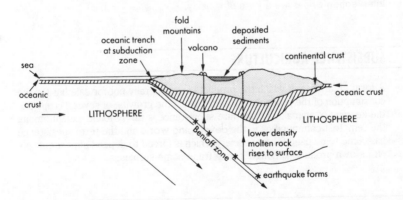

Fig. S.4 The destructive plate margin

towards the core of the earth along the **Benioff zone** where the subducting plate shears and descends along the edge of the continental plate. This can cause **earthquakes** and gives rise to a zone of volcanic activity on the surface above as lower density molten rock rises to the surface, see Fig. S.4.

SUBGLACIAL MORAINE

◀ Moraine ▶

SUBSERE

A stage in the evolution of a vegetation community towards a climax when the succession has been halted by non-climatic factors which are topographic, edaphic (soil) or biotic. In a dune system an example of a subsere would be a dune slack plant community which remains because the topography of the dune system does not alter to remove the slack over a period of years.

◀ Sub-climax ▶

SUBSIDY

A form of support given to farmers to encourage them to grow a crop such as oil-seed rape for vegetable oils in the European Community. Subsidies may also be given to support marginal farming and to help maintain the population in an area. Support for crofting in Scotland is in the form of subsidies. Subsidies may be political as in Switzerland to ensure agricultural production in the event of European wars isolating the country from inports of food. Intervention prices are a form of subsidy.

SUBSISTENCE AGRICULTURE

A form of farming in which the products are generally not for sale but for the consumption of the producers. It depends on the growing of several crops and the keeping of animals. Today pure subsistence is rarely found except among the remotest communities in the developing world and the term subsistence has come to be used for any society which is forced to depend almost entirely on its own produce although it may trade some surpluses.

SUBSTITUTION

The use of an alternative material to replace a resource at or towards the end of its cycle of resource exploitation. Aluminium cans are replacing tin cans so conserving the reserves of tin for more essential uses than Coca Cola! Other substitutes include glass fibre optic cables instead of copper cables and plastics instead of metals.

SUBURBANISATION

The process of city growth involving the decentralisation of people, jobs and activities to suburbs. The process began with building beyond the city walls in extra-mural suburbs although it is more usually associated with the spread of the city in the late eighteenth and early nineteenth century. The most rapid expansion of suburbia took place in the 1919–39 period and resulted in measures being taken to control the spread of cities – see **green belt**. It is the first stage in the **cycle of urban development**.

SUCCESSION

1 The second stage in the ecological process of **invasion** and succession which alters the patterns of social groups in residential areas by displacing the existing social groups.
2 The changes in a plant **community** which take place as that community moves towards the stage of a climax. This is also known as a sere where the initial community or **prisere** develops towards a climax such as a **psammosere** or sand dune community.
3 The sequence of strata of rocks at a particular place.

SUNRISE INDUSTRY

The type of company with a product or products at the early stages of the **product life cycle** who locate in **newly industrialising regions**.

SUNSET INDUSTRIES

The type of companies whose products are associated with the latter stages of the **product life cycle** which are normally located in declining industrial areas of **deindustrialisation** regions.

SUPERMARKET

A self-service store selling predominately foodstuffs, developed in North America it became common in Great Britain in the 1960s. It relies on low profit margins and a fast turnover of goods which attract customers who are prepared to buy all their requirements at one shop. The principle has led on to the development of the **hypermarket** and **superstore**.

SUPERSTORE

Often confused with and used interchangeably with **hypermarket**. It is a large store selling a range of goods, but not as large as a hypermarket. It is located on the outskirts of a town. Increasingly the superstore is becoming a specialist store catering for one type of purchase such as Do It Yourself goods, furniture and carpets, or electrical goods.

SUPRAGLACIAL MORAINE

◀ Moraine ▶

SURFACE STORE

A stage in the **hydrological cycle** in which water remains on the earth's surface in puddles either on the soil or on the roads and buildings. The water is either evaporated or eventually percolates the soil once the **infiltration capacity** is no longer exceeded.

SURFACE WATER GLEY

A soil showing features associated with the process of **gleying** in the surface horizons. This is a result of water logging at the surface normally associated with flooding derived from surface water flow. (See Fig. G.3.)

SURVIVAL LEVEL

◀ Optimum population ▶

SUSPENDED LOAD

The material carried by a stream in suspension, normally the smallest particles. However, in periods of **flood**, larger particles can be suspended in the stream because it has a greater velocity.

SUSPENSION

A method by which material is **transported** by streams, glaciers, the wind and waves. It is the most important method of transportation of lightweight sediment. The size of the particles in suspension is dependent on the velocity of the stream or wind, except in the case of glaciers.

SWALLOW HOLE

The widened area where two **grykes** intersect where water may flow down into the rock below to enter an **underground cave system** which is better known as a **sink hole**. The point is where a stream enters an underground system and, therefore, leaves a **dry valley** leading downstream from the swallow hole which is at a higher level. The valley leading to the swallow hole is said to be blind, i.e. it has eroded downwards faster than the rate of weathering beyond the swallow hole where there has been less river erosion.

SYSTEM

A set of logical procedures resulting from or acting upon a set of inputs which lead to the production of outputs. The procedures or throughput are capable of maintaining the system in operation or of altering or transforming it. Environmental systems are open systems because they depend on inputs of energy and matter and produce outputs of energy and matter. The concept of the system is an abstraction of the real world but it does enable the complicated processes taking place to be more readily comprehended.
◄ Catchment basin system ►

TAAFFE, MORRILL, GOULD MODEL

A model which outlines the links between transport and economic development through six stages as shown in Fig. T.1:

  a) scattered ports and trading posts along a coast with limited hinterlands;
  b) transport development favours selected ports which expand their hinterland;
  c) nodes develop on the feeders and further advantage is gained by the larger ports;
  d) prime routes develop further nodes with their own feeders;
  e) interconnection is complete as all the links are made and the nodes are arranged in a hierarchy;
  f) some routes carry more trade as main streets and develop trunk routes between them.

The development of railways in Liberia followed this pattern.

A   scattered ports

D   Beginning of interconnection

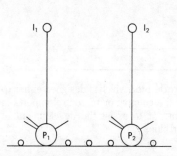

B penetration lines and port concentration

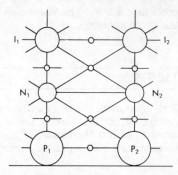

E complete interconnection

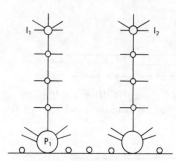

C development of feeders

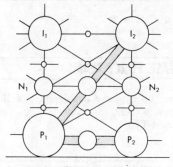

F emergence of high-priority 'main streets'

Fig. T.1 The Taaffe, Morrill and Gould model

TAKE-OFF

◄ Rostow's stages of economic growth model ►

TALUSFOOT

◄ Slope form ►

TARGETTING

The policy which attempts to make assistance for economic development more spatially and sectorally focused. As a result, regional policy has been replaced by urban policy and initiatives deliberately focused on a particular problem or activity. The **Merseyside Task Force** of 1981–3 was an example of this policy in practice.

TARIFFS

A tax upon goods entering a country or **trade bloc** which is designed either to protect the industries and agriculture of the country or to control imports in order to protect the balance of payments or to assist the development of economic activities within the country or bloc.

TARN

◀ Cirque ▶

TASK FORCE TEAMS

Set up by the Confederation of British Industry (CBI) since 1987 to aid development in named cities such as Birmingham and Newcastle-upon-Tyne.

TECHNOLOGICAL TRANSFER

The movement or sale of technologies for agricultural and industrial development to **developing countries**. It is a strategy used by **transnational corporations** to shift routine production to the lower cost sites in the developing world while, at the same time, claiming that they are assisting the developing world's development Some developing countries insist that they share in the equity of the plants established to utilise the transferred technology so that they gain more than just the technology. Both Singapore and Malaysia use the strategy of shared equity in industries and in tourist developments.

TELECOMMUNICATIONS

The technologies of rapid communication of information which include the telephone and its derivatives such as teleprinters and fax machines. Television-based communications systems often using computer-based data banks such as Prestel and Ceefax are other means of telecommunication. The development of telecommunications has influenced the type of buildings used by **producer services** and their location.

TEMPERATE DECIDUOUS BIOME

The natural vegetation of much of Europe, western North America, the eastern seaboards of N America and Asia and small areas along the Chilean coast and eastern Australia and New Zealand. This biome of deciduous trees with some needle leaved trees, generally has broad leaves with some climbers such as ivy and epiphytes. The number of species is low and its productivity is 66% of that of the **tropical rain forest**. There is a shrub layer present because light can penetrate. Nutrients are stored in the soil and vegetation because litter is decomposed efficiently. The soils are **brown earths**. This biome has been most adapted by people, so much so that very few areas of it exist. Remnants are found as small woodlands and forests or in hedgerows. **Coppicing** was a method of increasing its productivity. In the past century much of the forest has been replaced by faster-growing coniferous trees in the interest of increased commercial exploitation of the land. This is often opposed today because of the changes which it brings to the landscape and its appreciation by people. Therefore deciduous forest is now being replanted for aesthetic reasons. Some areas are reverting to woodland because they are no longer needed for agriculture and because constraints on tree and shrub growth have been removed, e.g. rabbits were killed by myxamatosis and so new shrubs and saplings could grow, e.g. the South Downs escarpment.

TEMPERATE GRASSLAND BIOME

A grassland biome with some trees found in the heart of the northern continents including the **steppes** and **prairies**, the **pampas** of S America and the **veld** of South Africa. The grasslands are grazed by large herbivores. Nutrients are stored predominantly in the soil which is high in humus, **black earth** or **chernozem**. Some **ecological niches** exist for trees such as aspen. People have adapted the area by firing it, by hunting to replace, e.g. bison, with domesticated herbivores and by ploughing up for cereal cultivation which eliminated insects and plants in the simplified **ecosystem**.

TENSIOMETER

A porous, permeable ceramic cup connected through a tube to a manometer or vacuum gauage. The cup is in contact with the undisturbed soil at a specific depth and thus the instrument measures negative pressure or soil water tension *in situ*.

TEPHRA

◀ Volcano ▶

TERMINAL MORAINE

◀ Moraine ▶

TERTIARY SECTOR

The sector of the economy involved with the distribution and retailing of manufactured goods and primary products including a number of professional and personal services and public administration. It can be called the **service** sector which is then subdivided into **producer** and **consumer services**.

TGV (TRAIN A GRANDE VITESSE)

The French high speed passenger train which already runs to Lyon and to the west coast and will soon run to Calais, Belgium and West Germany. It is proving to be a competitor for the airlines over distances up to 500 km.

THERMAL POLLUTION

Pollution caused by the waste heat from power station cooling systems and factories which is discharged into streams and lakes increasing the temperature of the water. Warmer water increases the demand for oxygen and the lack of oxygen will kill some species.

THIRD WORLD

◀ Developing world ▶

THREE FIELD SYSTEM

The dominant agricultural practice in mediaeval times in which arable land was divided into three open fields or areas of cultivation. Each one of the fields would lie fallow for one year, be cultivated the next and be used as meadow land in the final year of the rotation. The system enabled the fertility of the land to be maintained although it was expensive of land. The system disappeared when enclosure, the abolition of common rights to land and the changing of the land to freehold with boundaries in the form of fences and hedges, took place mainly between the 16th and 18th centuries.

THRESHOLD

The minimum demand or population needed to support the provision of a good or service.
◀ Christaller's central place theory ▶

THROUGHFALL

Part of a stage in the hydrological cycle where precipitation stored on plants falls as drips either to lower plants or to the ground surface. It is usually paired with stemflow, the movement down tree trunks and plant stems from the interception store to the ground surface.

THROUGHFLOW

The lateral movement of water within the soil regolith, i.e. between the bedrock and the surface of the soil. It is a lateral movement because not all the water can percolate downwards as the regolith is not permeable enough, e.g. there is a hardpan. It results in the downslope movement of water and also moves fine soil particles downslope. It also plays a part in the pedological process of eluviation known here as lateral eluviation.

TIDAL POWER

◀ Energy resources ▶

TIDES

The rise and fall of the level of the oceans and seas caused by the gravitational attraction of the moon and sun. A consequence of this regular rise and fall is the tidal current, the movement of water in and out of an area (bay, estuary or channel). The tidal range is the difference in the height of the water between high and low tide. It varies according to the strength of the attraction between the sun and moon, i.e. between spring and neap tides. Tidal range increases in estuaries and bays where the water is pushed into an increasingly confined space such as the Severn Estuary and the Bay of Fundy. It is this range which is exploited by the **tidal power** station on the River Rance in Brittany and is proposed to be expoloited on the Severn and in Morecambe Bay.

TIED AID

◄ Aid ►

TILE DRAIN

A subsurface pipe of concrete or ceramic material used to conduct drainage water and avoid water logging.

TILL SHEET

◄ Drift ►

TOPOLOGICAL MAP

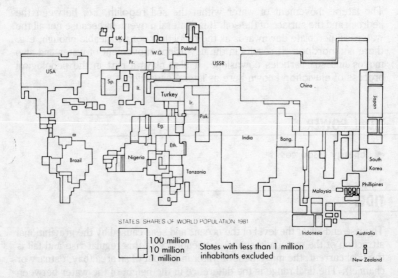

Fig. T.2 Topological map of world population in 1981

A map such as that in Fig. T.2 drawn so that the areas are proportional to the size of the phenomenon being measured. The shape of the area is adjusted to make it approximate to the real world. Topological maps are used to show, e.g. the dominance of population numbers in the developing world and other aspects of unequal development.

TOTAL POTENTIAL

A measure of the amount of work to be done to move a unit quantity of pure water at a specified elevation in a soil:

$$\psi t = \psi c + \psi g + \psi \pi + \psi p$$

- Capillary potential; (c)
  force resulting from the pore size structure of the soil.
- Gas pressure potential; (g)
  force resulting from differences with respect to external gas pressure.
- Gravitational potential; (π)
  force resulting from the position of the soil site with respect to the gravitational field.
- Osmotic potential; (p)
  force resulting from differences in salt content of water.

TOURISM

Recreation which normally takes place away from the home base. It is a more commercial form of leisure because it involves overnight stays in accommodation such as hotels and because it involves foreign visits and flows of invisible earnings between countries.

TOWN AND COUNTRY PLANNING ACTS 1947, 1968

◀ Town planning ▶

TOWN DEVELOPMENT ACT 1952

Set up the machinery for the development of expanded towns whereby local authorities in an overcrowded metropolitan area made agreements with an authority elsewhere to make land available for the housing of overspill population and industry. These arrangements were made most often by London and resulted in the development of Swindon, Basingstoke and Haverhill.

TOWN PLANNING

In Britain the main foundation for town planning was laid by the 1947 Town and Country Planning Act which established the principles of development control and land use zoning. This act was superseded by the 1968 Town and Country Planning Act which introduced the threefold division of planning into **structure plans, district plans** and **local plans**. Structure plans have to be prepared for each county and are a broad-brush approach to the development needs of a county for a period of years, normally a decade. It provides in its

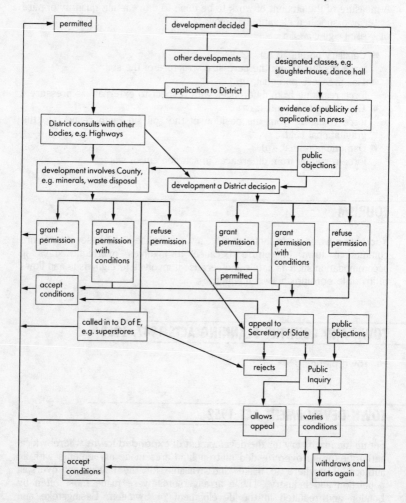

Fig. T.3 A simplified flow diagram of the development control process

written statement and accompanying maps a framework for district and local plans. **Public participation** is an essential part of the process of plan preparation. District plans must all conform to the broad guidelines of the structure plan. They are comprehensive plans for an administrative area designed to provide the framework for both large and small scale changes. Action area plans are short-term plans for a small area or a specific goal such as the rehabilitation of a former industrial site. Subject plans deal with specific uses such as marina development or retail warehousing. All plans are part of the **development control process** which involves the participation of developers and planners. If a development is refused the developer may go to appeal and whoever loses the appeal pays the costs. Counties have to assume that all development is in the public interest unless they can prove otherwise. For this reason alone, planners have sometimes been reluctant to refuse permissions. Any building development, whether beneficial to the area in the long term or not, is lucrative to the local authority, and the loss of this potential revenue added to the danger of legal costs from a successful appeal may cost the ratepayers dearly.

Changes of use of buildings especially to retail and office use from industrial use has been made easier and so planning control has been relaxed in favour of the developer as a result of the **Local Government and Planning Act 1980**. In 1990 consideration was being given to the abolition of structure planning so eroding further the broad regional interest in planning. The stages are shown in Fig. T.3.

TRACTION

A method by which eroded material known as **bedload** is transported along the bed of a stream. The greater the velocity the bigger the traction load and therefore most traction takes place during periods of flood.

TRADE BLOC

Regionally based, international organisations such as the **European Community** (EC), LAFTA (Latin American Free Trade Association and ASEAN (Association of South East Asian Nations) which have been established to encourage the growth of trade between their members partly by erecting external tariffs to enable production to grow internally. Such measures can restrict trade although some argue that the growth of an internal market can stimulate trade. Sometimes blocs make agreements with other countries so that they may gain preferential access to the market, e.g. Lomé Convention which gives preferential access to the European Community for the former colonial territories in Africa. Since 1948 GATT (General Agreement on Tariffs and Trade) has met regularly to try to eliminate the effects of **protectionism** by trade blocs which are a barrier to free trade. Since 1964 UNCTAD (United Nations Conference on Trade and Development) has also tried to reduce the effect of trade blocs on developing economies.

TRADE WINDS

The winds associated with the Hadley cell originating in the **tropical continental air masses** to the eastern side of the major oceans and blowing towards the equatorial low pressure **(ITCZ)** as north east trades in the northern hemisphere and south east trades in the southern hemisphere. The term originates from the era of sailing ships which made good use of the relatively constant wind speed and direction to speed their passage.

TRADITIONAL SOCIETY

◀ Rostow's stages of economic growth model ▶

TRANSFER/TRANSLOCATION

Two terms used to describe the reorganisation of material and energy within the soil profile. They describe the movement of material (e.g. clay, humus, iron, etc.) and soil water up and down the soil profile. They are a process of **pedogenesis**.

TRANSFORMATION

A process of reorganisation of material and energy within the soil profile that does not involve movement, i.e. *in situ*. For example, the change of primary mineral into secondary mineral by weathering. A process involved in pedogenesis.

TRANSFORM FAULTS

◀ Plate tectonics ▶

TRANSHUMANCE

The seasonal movement of animals in mountainous areas up to the higher pastures in summer and down to the valley in winter. The particular morphologies of glaciated valleys have lent themselves to this practice because the high pastures are situated on **truncated spurs** and in **hanging valleys**. Today this practice, which attained ritualistic significance in their communities at certain times of the year, is in decline due to less dependence on locally produced food in many mountainous regions. Nevertheless, the animal migration times do form a tourist attraction.

TRANSNATIONAL CORPORATION

Used interchangeably with **multi-national company** although it is the preferred term of the United Nations. Its productions of goods and services comes from a number of countries. The company's interests are not focused upon any one country and control is not necessarily in one country. The transnational companies are able to operate in a way which enables them and their products to avoid tariff barriers and to manufacture in those countries where the capital and labour costs are lowest.

TRANSPORT

1 The carrying of goods and/or people between places. It is sometimes called transportation.
2 In geomorphology transport refers to the movement of material from the place of **erosion** to the site of **deposition**. Transport by water is by the processes of **saltation, solution, suspension** and **traction. Waves,** currents and **tides** transport material in marine environments. Material is also transported by the wind by saltation, suspension and traction; glaciers transport material in suspension. Gravity is another agent of transport. While in transport the materials or **load** are further eroded by **attrition** and themselves **abrade** the surface. During transport the load is **sorted** and becomes graded which can be seen in most deposited materials.

TRANSPORT GEOGRAPHY

The study of the role of transport in the economic geography of an area. It focuses primarily on the patterns and modes of transport although specific studies of **networks** and transport terminals have become important. Transport is also studied as an impetus for change as new methods after locational choice, e.g. **containerisation**. Transport is also seen as a social need and so studies of access to transport for, e.g. the elderly, are important.

TRANSPORT MODES

The different means of carrying goods and people such as rail, inland waterway, pipeline, ocean vessel, road vehicles and aircraft.

TRICKLING DOWN

◄ Growth pole theory ►

TROPHIC LEVEL

The point at which energy is transferred from one plant or animal to another as a part of a food chain. The **primary producer**, grass is the first trophic level which is consumed by a rabbit, a **herbivore**, which, in turn, as the second trophic level, is eaten by a fox, a **carnivore**. At each stage energy is lost to the system so that only 10% is passed to the next trophic level.

◄ Ecosystem ►

TROPICAL CONTINENTAL AIR MASS

Air originating over the major hot desert regions of the world beneath the Hadley cells. It rarely reaches the UK although, when it does, the weather conditions are very mild in autumn, winter and spring and very warm to hot in summer.

TROPICAL MARITIME AIR MASS

Air which originates over the oceans in the region of the tropics; in the UK this is the Azores high. It brings mild conditions in winter and warm weather in summer. It is very **stable**.

TROPICAL MONSOON CLIMATE

A climate based upon the interaction of large scale, planetary circulation and more local factors such as the relief of the area as shown in Fig. T.4. The

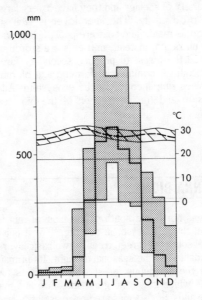

Fig. T.4

climate is markedly seasonal (*mausin* from which 'monsoon' is derived is Arabic for season). The winter season is dry and associated with offshore winds caused by subsiding air from the tropical jet stream, whereas in summer the wind pattern is onshore with winds being drawn inland by strong thermal low pressure systems which migrate inland as the jet breaks up or migrates polewards. The advent of the monsoon is associated with heavy rains. The pattern of land and sea areas and relief will affect the detailed pattern at any place. For example West Malaysia is affected by the summer monsoon from the south west whereas the east coast of West Malaysia is in the rainshadow and is more affected by the winter monsoon winds which have crossed the South China Sea and bring heavy rains to that coast. The pattern of the seasons affects the cropping patterns and routines and the timing of the main tourist seasons.

TROPICAL RAIN FOREST BIOME

Found in 3 main areas; the Amazon and Zaire basins and SE Asia. Vegetation is characterised by being evergreen with leathery leaves, buttress roots and cauliflower-like flowers on the trunks. Epiphytes are plants which grow on the trunks of the larger plants and lianas or climbers grow from the ground to the trees. Saprophytes are the plants which grow with little light and little chance to photosynthesise, depending mainly on rotting vegetation; the large fungi found in the forest are saprophytes. Ground level vegetation is minimal. There is a great variety of species and they have a very large **biomass** with a high **net primary productivity**. The **litter layer** is well supplied from the vegetation and decay is rapid. The soils are poor, ferruginous soils, **latosols**, which contain few nutrients, instead, nutrients are stored in the vegetation. People have adapted the forest to produce secondary forest as a result of slash and burn agriculture, **commercial farming** and **plantation agriculture**. All adaptation involves **simplification** of the ecosystem. Adapation may also result in **soil erosion** and the destruction of the biomass leading to the **greenhouse effect**.

TROPICAL SAVANNA BIOME

Found in those areas north and south of the tropical rain forest where the rainfall pattern is much more seasonal, see Fig. T.5. It is a grassland biome with tall grasses and some evergreen trees in two basic layers. The plants are xeromorphic, i.e. adapted to the seasonal drought. Its primary productivity is about the same as the deciduous forest which provides food for the large **herbivore** mammals which graze the grassland. Fire is a major factor in the control of the biome and many species are fire-resistant; it is probable that this

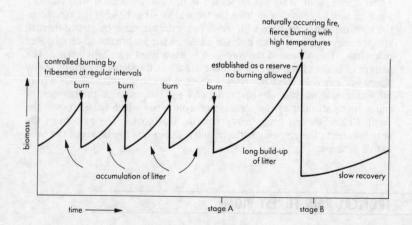

Fig. T.5 The effect of burning on an area of savanna vegetation

is not a climax vegetation. The biome is essentially one found on plains and plateaux. Adaptation has come as a result of fire favoured by herders to increase nutrient levels and to replace the sparse woodland. People have put pressure on the biome with the result that **desertification** has been exacerbated by the need to increase herds of goats and cattle and collect more timber for fuel to support the growing populations of these areas. The charity Bandaid was set up to help combat the combined effects of **drought** and the human consequences of overexploiting this biome. Agriculture has **simplified** the ecosystem as well.

TRUNCATED SPUR

A valley side spur which has been cut off by a valley glacier so leaving a steep, cliff-like face in place of the gentler slope of the spur. The removal of the spurs gives the glacial valley a much more straight and U-shaped appearance.

TSUNAMI

◀ Earthquake ▶

TUNDRA BIOME

The biome associated with the **periglacial** areas polewards of the **boreal forest biome**. The southern limits of the region are defined by the tree line or more accurately by areas where the mean daily temperature for the warmest month does not excede 10°C. The plants are adapted to the extreme cold and short growing season so that they grow rapidly in the long daylight hours of summer and include many berry-producing varieties whose seeds are dispersed by birds. The leaves are small and evergreen and the shrubs are very woody. Mosses and lichens are abundant because they can photosynthesise at temperatures as low as −20°C. Dwarf shrubs and trees characterise the southern margins of the biome. The foodchains are simple with the **herbivores** migrating in winter. The system has been affected by the hunting of both herbivores such as caribou and the **carnivores** such as bears.

UNDEREMPLOYMENT

The partial unemployment of workers in rural areas especially in the developing world where there is not enough land to occupy the whole population. It can be regarded as a product of overpopulation.

UNDERGROUND CAVE SYSTEM

The caves which develop within an area of carboniferous limestone whose locations are related to the position of the water table at various stages in the formation and evolution of the cave system. Caves may be phraetic, that is water filled, or vadose, that is containing an underground stream. They are therefore a hidden feature of karst scenery.

UNDERPOPULATION

Occurs in an area when an increase in population would result in the more effective use of resources which would also raise living standards. The population level is below the optimum population.

UNEMPLOYMENT

The involuntary lack of work which is normally expressed as the number who want and are seeking work as a percentage of those in work. It can be subdivided according to sex and age because the patterns differ and have different causes. It is frictional if it is caused by people changing jobs and structural if it is caused by a mismatch between the available skills of the workforce and the available jobs. Much unemployment in the North East is structural due to the decline of shipbuilding work. It is the product of long term shifts in employment such as deindustrialisation, technological developments in industries and the substitution of new products from new areas for old products from older industrial areas. Unemployment may be concealed because the government excludes certain categories of unemployment; this occurs in Great Britain where various employment schemes of a temporary nature keep people off the unemployment register.

UNITED STATES TYPE

◀ Population-resource ratio ▶

UNTIED AID

◀ Aid ▶

URANIUM

A radio-active metal normally found in igneous rocks, shales, limestones and sandstones which is the source of isotopes used in the production of nuclear energy for **atomic power.**

URBAN GEOGRAPHY

The study of urban areas has developed since the 1950s to be a major area of geographical analysis. It is both the study of settlements in a country or region, i.e. the urban system – see **Christaller** – and the study of the arrangement of activities and functions within the city. Therefore urban geography contains many other specialisms such as retailing **offices,** housing, **social geography, transport** and **political geography.** There is often an **applied** aspect to urban geography because many of the studies are of direct use to the urban planners and to the users of the city. Studies of the impact of retailing developments are just one example of applied urban research.

URBANISATION

The process of becoming urban. It was a term devised to convey the rapid growth of urban populations especially during the industrial revolution in Britain. Settlements grew as a result of **rural to urban migration** and rapid **natural increase** due to the youthful population structure. It is also a social process because it results in the development of a more stratified society with a class system. It took place in different countries at different times. Today urbanisation in the **developing world** is different from European and American urbanisation. Cities still attract people to them but the speed and scale of urbanisation is much greater. People are not pulled by jobs as they were a century ago but rather by the *prospect* of a job so that urbanisation in the developing world is often seen as urbanisation without **industrialisation.** People have migrated on the basis of hopes and dreams. **Underemployment** is often the push in rural areas which is transferred to **unemployment** in urban areas. Urbanisation may take place as **new towns** in the socialist world.

URBAN MORPHOLOGY

The arrangement of land uses within an urban settlement which takes account of its growth and shape as well.

URBAN PROGRAMME

The first initiative announced in 1968 to combat the **inner city problem**. It was cut back in 1983–4 as the **enterprise zone** policy was extended.

URBAN–RURAL MIGRATION

◀ Counter–urbanisation ▶

URBAN SETTLEMENT

A settlement containing a range of functions which increase in number with increased size of population. It is normally seen as a settlement of over 3,000–5,000 population which is the upper limit to a **village**. The minimum size varies according to the official definition of many countries. Towns and cities are two levels in **Christaller's central place theory**.

URSTROMTÄLER

◀ Meltwater ▶

UTILIDOR

A heated, insulated, above ground conduit for utilities such as water supply, waste disposal, gas and electricity used to counteract the problems of **permafrost** in the Northlands of Canada. Similar conduits exist in Northern Scandanavia and the USSR.

VALLEY TRAIN

An accumulation of **fluvioglacial** deposits down a valley for a great distance from the ice front. These areas have been subsequently reworked by post-glacial fluvial erosion so that the area of deposition might not be continuous.

VARVED CLAY

◀ Proglacial lake ▶

VELD

◀ Temperate grassland biome ▶

VELOCITY OF FLOW

◀ Channel variables ▶

VENTIFACTS

Stones in a desert which have been polished by wind-blown sand.

VERTICAL LINKAGE

A concept in industrial geography used to refer to firms who are all linked together in that the product of one is the raw material of the next plant. It was best applied to the textile industry in the past where spinning firms supplied weavers, etc.

VICIOUS CIRCLE OF POVERTY THEORY

Nurske showed that poverty is self-perpetuating and that the **developing world** countries are trapped by their lack of capital and low incomes as shown in Fig. V.1. The theory assumes that development takes place in isolation and

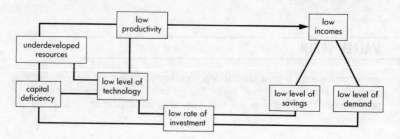

Fig. V.1 The vicious circle of poverty

that **aid** and loans play no part in development. It is possible to break out of the circle as the **newly industrialising countries** have shown, e.g. Singapore and as communist China has also shown to a more limited extent.

VILLAGE

A form of rural settlement normally containing a church or religious building and some low order functions: they are classified according to shape. Linear villages are found along valleys, dykes or a road. Nucleated villages are found at natural focuses, junctions, defensive points or where there are certain physical conditions such as a spring line. Villages can also be classified according to their morphology, e.g. green, *strassendorf* (street village), planned village. The population of a village is normally between 200 and 3,000 with some planners saying that a village cannot support sufficient low order functions unless it is as large as 5,000 people. 'Village' can be used by the property development business to signify any small development and many centres of old villages now swallowed up in suburbia may be perceived as villages by the residents. The terms **metropolitan village**, dormitory village and **commuter village** are used to describe those settlements which are within commuting range of major settlements and populated by commuters.

VISIBLE TRADE

The flow of both **primary produce** and **secondary** manufactured goods from producers to market measured in terms of the cash value of that trade. Trade is dominated by the **developed world**.

VOLCANO

The major landforms produced by **extrusive vulcanicity**. Volcanoes when active have two types of eruption; **hawaiian** which is the less violent and results in the emission of basic lavas from the oceanic crust which flow very easily and **plinian** which is more violent and forms a large cone from its slower moving, acid lava, from the continental crust. The plinian is always more explosive, e.g. Vesuvius and Mount St Helens. Some eruptions have been so violent that they have created a new, larger crater replacing the former smaller cone; this is how Crater Lake in Oregon formed in the **caldera** caused by the explosion. Volcanoes are a hazard in that they cause destruction to large areas, leave **scoria** or ash dust, **lapilli** or small solidified stones thrown through the air, and **tephra** or rock fragments thrown out as molten lava which solidify as they pass through the air, (the term **pyroclastic material** covers all material thrown into the air by eruptions). **Pumice**, a sponge like rock formed from the scum on the surface of a lava flow or blown out from an eruption is also produced. Volcanoes are also a benefit to people. The soils derived from the lava are often fertile and many volcanoes are a tourist attraction. Nevertheless it is the threat to livelihood which normally makes the headlines in the short term.

VON THÜNEN'S MODEL OF AGRICULTURAL LOCATION

Von Thünen was a Prussian landowner (not a geographer!) who published his theory of agricultural location in 1826 entitled 'The Isolated State'. His aim was to explain the general principles governing the prices of farm products and the way that these prices controlled the agricultural product on any piece of land. His assumptions were; one market, one mode of transport and that farmers supply one market and respond to that market, all very much the conditions which existed at his time. There were three principles:

- the price obtained by a farmer declined with distance from the market because that final price was the selling price minus the cost of transport;
- the nearer a farm to the market the higher the returns because competition for the land raised its price and therefore forced farmers to make the most profitable, intensive use of the land;
- **economic rent** for each crop fell with distance from the market which is why there are concentric rings on the model shown in Fig. V.2. The economic rent for each crop diminishes at a different rate from other crops.

He went on to show how changes in market demand will alter the pattern of farming around the market and thus his model is one of agricultural change and not just of static land use rings.

The theory has been applied to several locations at a continental level (Europe in the 1930s), national level (Uruguay), regional scale (New South

Wales) and local level (villages in N Nigeria). As von Thünen forecast, the factors modifying the pattern of land uses have changed:

- there is more than one market;
- transport modes have changed so that produce can be grown where the conditions are most suitable and then refrigerated for transport to the market;
- new crop strains enable crops to be grown in new areas;
- political and social influences are more important today;
- the land surface is not **isotropic** with the same conditions everywhere;
- the patterns have changed so that grain farming is found further from markets as are certain types of dairying.

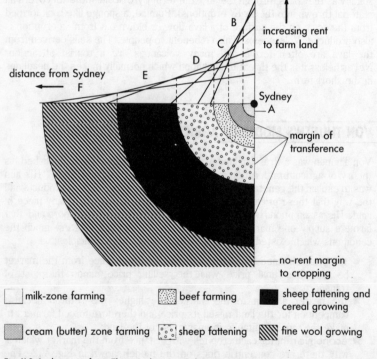

milk-zone farming

beef farming

sheep fattening and cereal growing

cream (butter) zone farming

sheep fattening

fine wool growing

Fig. V.2 Applications of von Thünen's theory to Sydney, Australia

VULCANICITY

◀ Extrusive vulcanicity, Intrusive vulcanicity ▶

WADI

A valley in a desert or semi-desert environment whcih has been formed by running water. The valley is normally steep-sided as a result of the occasional rapid erosion. Such a valley may be permanently dry having been formed during wetter times in the past or may contain an **ephemeral** (intermittent) **stream** after intense periods of precipitation.

WANING SLOPE

◄ Slope form ►

WATER DEFICIT

A situation in which annual potential evapotranspiration exceeds annual precipitation. It is normally associated with desert areas although it is possible for many temperate areas to have a deficit over a year, e.g. 1989 in southern England.

WATERLOGGING

A state of the soil when the system is reduced to one of only solid and liquid, i.e. all pores are filled with water.

WATERSHED

The boundary of a catchment basin where precipitation falling on either side of the watershed will enter two separate **catchment basin systems**. It can refer to the whole of the catchment system.

WAVE CUT PLATFORM

A level area on the coast extending between the high and low water marks which has been cut by the waves as the cliff retreats. Normally the top of the platform will be covered in debris fallen from the cliff face which may merge with the **storm beach** and sometimes the platform may be covered in shingle and debris which will be removed by storm conditions.

WAVELENGTH

◄ Waves ►

WAVE POWER

◄ Energy resources ►

WAVES

In hydrology and oceanography it is the product of wind presssure on the water surface which causes the surface to bend in a regular pattern. The waves have an amplitude, i.e. a vertical distance from the trough to the crest and a **wavelength**, the distance between two crests. The size of the waves is proportional to the wind speed generating the waves, the duration of that wind and the **fetch**, the distance of open water over which the wind was blown. Waves are the principal agent of marine **abrasion** although they also erode by **hydraulic action**. Waves also **transport** eroded material by the process of **longshore drift** and through the action of **swash** and **backwash** move and deposit sediments along a coast. Waves may also be **seismic** or shock waves, i.e. generated by an earthquake and recorded by seismographs. Waves formed by submarine earthquakes are known as **tsunami**.

WAXING SLOPE

◄ Slope form ►

WEATHERING

The process by which rocks in the earth's crust and the **regolith** are broken down and/or decay. It can be subdivided into **chemical, mechanical** and **organic weathering**. It is the first stage in the process of **denudation**.

WEBER'S MODEL OF INDUSTRIAL LOCATION

A model of industrial location devised in 1909 and based on transport costs. If the costs of transporting the raw materials are the same as the cost of transporting the good to the market then the plant will be located midway between market and the resources. If the materials are cheaper to transport than the finished product then the location will be nearer the market and the industry is weight gaining. If the raw materials are more costly to transport, the location will be nearer the source of the materials and the industry is weight losing. Weber illustrated these principles graphically (see Fig. W.1)

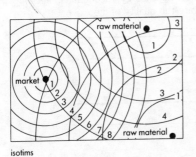

isotims

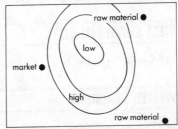

isodapanes

Fig. W.1

using lines of equal cost from the market and the raw materials called **isotims**. The spacing of these lines or contours of costs varies according to the costs of each form of transport. These costs are aggregated at points on the surface (**isotropic** plain) to enable one to construct **isodapanes** or lines of equal transport cost. The area within the lowest cost isodapane is the area with the least cost location.

The model assumes that there is no preferred transport route and that there are no barriers to transport on the isotropic surface. He also assumed that transport costs rise with distance but in fact they can vary due to loading and unloading costs particularly where different modes of transport are added. The model ignores the role of government in pricing policies. Transport costs only make up a small proportion of total costs (5–12%) and so other factors are more important. Weber concentrated on cost minimisation and ignored the role of revenue in a firm's choice of location. He also undervalued the role of **agglomerative** and **deglomerative** forces in location decisions. The theory pays no attention to behavioural factors, such as knowledge of all locations when decisions are made. The structure of the firm will also influence the choice of location. Weber's theory is thus a starting point for industrial location studies which really reflects the main types of location of industry found at the time when the model was first devised.

◄ Lösch's theory of industrial location (1954) ►

WELFARE GEOGRAPHY

This is an approach to human geography which emphasises inequalities in society and in the world; it is concerned with the quality of life. It asks who gets what, where and how; in other words, what benefits are obtained where and by what processes they are obtained. It also asks about the disbenefits or losses in the same way. Such questions are worth asking of any planning scheme or any environmental conflict. Such an approach to geography has broadened to look at **hazards**. It is also an ideal way of indicating the interdependence of disciplines such as psychology, economies, sociology, geography, etc. today as we investigate the quality of life.

WEST EUROPEAN CITY

◀ Nellner's model ▶

WETTED PERIMETER

◀ Channel variables ▶

WILDLIFE AND COUNTRYSIDE ACT 1981

The legal basis for nature **conservation**. It enables the Nature Conservancy Council to designate **sites of special scientific interest (SSSIs)**.

WILTING POINT

An undefined specific soil moisture percentage when plants show physiological signs of wilting.

WINDBREAK

The use of a line of naturally occurring or artificially planted vegetation to protect a crop from the effects of the wind. These breaks may also be planted to ensure that the downslope movement of cold air at night does not damage crops although poorly located windbreaks may pond back cold air in a form of frost hollow.

WIND DEFLATION

The process by which the wind picks up small particles of sand and silt from on top of glaciers and redeposits them elsewhere. Such wind-blown sand and clay from the ice sheets is known as **loess** or **brickearth** and is found across much of the southern edge of the North European Plain. It gives rise to fertile soils associated with major areas of cereal and sugar beet growing. Brickearth is the term used in southern England where large deposits are associated with the **brown earth** soils on which much of the **market gardening** of Sussex is founded.

WIND POWER

◀ Energy resources ▶

WORKING POPULATION

◀ Active population ▶

WREFORD-WATSON'S MODEL

This model, illustrated in Fig. W.2, is a recent attempt to update the classic models of **Burgess, Hoyt,** and **Harris and Ullman** and to reflect the realities of the North American city in the late twentieth century. It recognises the growth of suburban business centres, satellite towns, the impact of transport and the existence of rural wedges penetrating suburbia. Even this model fails to recognise the out-of-town retail developments of recent years.

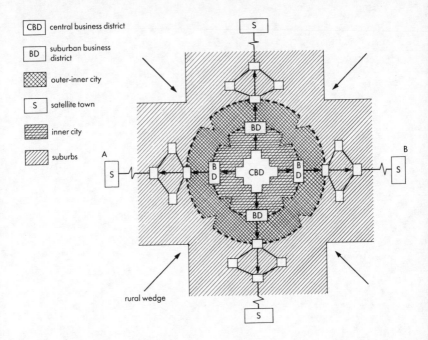

Fig. W.2 Wreford–Watson's North American city

YARDANG

A wind-formed ridge of rock in desert areas with a rounded windward face and a sharp ridge running down the leeward side. Each ridge is separated from the next by a wind-excavated groove.